I0488953

# Police Radar Handbook

## A Comprehensive Guide to Speed Measuring Systems

by

Donald Sawicki

POLICE RADAR HANDBOOK

© Copyright 2013, Donald Sawicki

All Rights Reserved.

# Contents

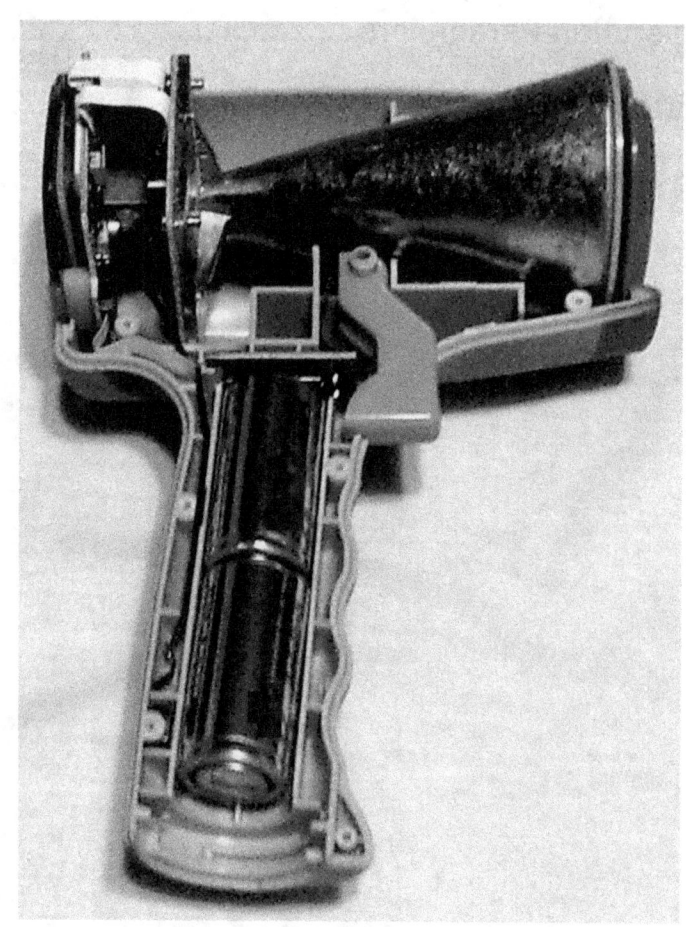

K Band Radar

# Preface

Police radar is as controversial today as it was when the 55 mph national speed limit started in 1974, a temporary law that lasted over 20 years. Around that same time solid-state police radars started showing up on the roads in large numbers. The advantage of solid-state replacing vacuum tubes was to bring down the cost, size and power requirements of the units. Cost reductions included *not* having a strip chart recorder for a written record of target history. In 1987 Congress raised the national speed limit to 65 mph and finally abolished it in 1995.

Even through the 55 mph national speed limit is history many communities continue to use, and some abuse, police radar. Many communities have become addicted to revenue obtained from radar. In the 1990's laser radar, also called lidar, started showing up on the roads. Police radar, both microwave and laser, is used extensively across the nation and around the world.

Many police departments and agencies have greatly improved their application of radar to traffic control, but there still exists a large number not qualified to operate this type of precision instrument. To operate a traffic radar does not require genius, but it does require proper training as well as a basic understanding of the device.

Vehicle Insurance Companies have an interest in police radar. Insurance companies base automobile insurance rates on driving record, the more speeding tickets a driver has the higher the insurance rate. In other words the more speeding tickets issued the more money the insurance companies make.

NOTES

# Chapter 1 - Speed Measuring Systems

## Chapter 1.1 - Police Traffic Radar

### Microwave Radar

radar (rã´där), noun. (1) acronym for **RA**dio **D**etection **A**nd **R**anging. (2) a remote sensor that emits electromagnetic radio waves, microwaves, or infrared laser light in order to measure reflections for detection purposes such as presence, location, motion, speed. (3) radiolocation. (4) field disturbance sensor. (**5**) proximity sensor.

Police microwave radars continuously transmit while simultaneously receiving and processing reflections. Reflections from moving vehicles are frequency shifted by the Doppler Effect and proportional to speed.  Police radars have minimum **sample times**, time required to process one speed measurement.  Sample times vary with model from about 1/4 to 1/3 seconds or more.  Typically multiple sample periods are required when an echo is first detected or if the radar has just started transmitting.

Police radars are available in 2 basic configurations with various options and modes.

### CONFIGURATIONS

**Hand Held Radar**      - Operates from a **stationary** position.

- Models with a moving mode also have a radar mount inside the patrol vehicle for moving mode operation.

**Fixed Mounted Radar**   - **Stationary** or **moving mode**.

- Some have optional second rear antenna.

### OPERATIONAL MODES

**Stationary Mode**   - Tracks **approaching** traffic.

- Tracks **receding** traffic. optional

**Moving Mode**      - Tracks **opposite direction** traffic.

- Tracks **same-direction** traffic. optional

*Cannot* measure traffic within **± 2 to 5 mph** of patrol vehicle speed.

Moving mode radar **requires a minimum patrol speed of 5 -20 mph.**

## TRANSMIT AND TRAFFIC MODES

**Transmit Modes** - Transmits continuously.
- Instant-on. optional
- Pulsed. *Unreliable.* optional

**Fastest Mode** - Tracks and displays 2 targets, the Fastest Vehicle
optional     and the Vehicle with the strongest reflection.

All radars track and display the strongest reflection.

### Beamwidth

Most microwave traffic radars have a relatively wide beam, 9° to 25°, that easily covers several lanes of traffic at a relatively short range. Detection range in the beam varies with radar and vehicle reflectivity and may be as low as 100 feet or less to 1 mile or more. A radar may track a distant large vehicle instead of a closer small vehicle without any indication to the operator which vehicle the radar is tracking.

### Cosine Effect Angle

The angle between the microwave or laser radar and target must be small for an accurate speed measurement. The angle is referred to as the *Cosine Effect* angle because measured speed is directly proportional to the cosine of this angle. The larger the angle the lower the measured speed. The angle also imposes a minimum range on the radar or lidar, vehicles inside minimum range are too close to measure. A radar in moving mode can measure speed high in some situations. **The radar should be located as close to the road as practical to minimize cosine effect errors and limitations.**

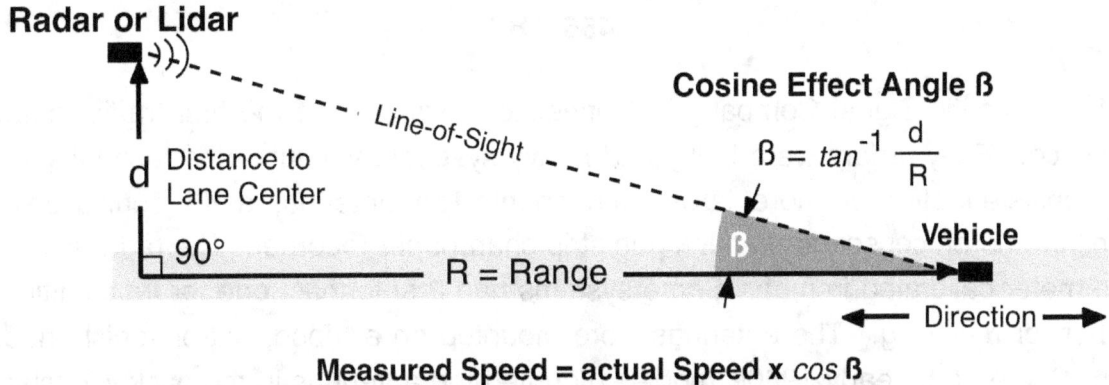

Figure 1.1-1 -- **Radar Cosine Effect Set-up**

## Timing Modes

Some microwave and laser radars have a timing mode that allows the operator to time targets. The operator measures the time it takes a vehicle to travel between 2 points of known distance. The radar is not transmitting rendering detectors useless. This method takes more time to set-up, requires more operator actions, is less versatile, and thus used less often.

## Dummy Radar

Some roads have unattended fixed mounted *dummy radars* that constantly transmit a signal in one of the Federal Communications Commission frequency bands approved for police radar. The FCC requires unattended transmitters to radiate less power than police radar. These transmitters do not have receivers and cannot measure speed or anything else. Dummy radars are intended to set off radar detectors to alert and fool drivers with detectors to travel at the speed limit. Locals usually figure out where the dummies are and ignore them, but be careful because police like to operate radar on a different frequency in those areas. A popular place to mount dummy radar is on overhead highway signs and mobile road electric signs.

## S Band Radar
### 2.455 GHZ

In 1947 Automatic Signal Company in Connecticut built one of the first traffic radars for state police.  Early radars were bulky and heavy systems, vacuum-tube technology.  The radar consisted of 3 or more large components, an antenna, a 45 pound box with transmitter and processor, and an ink pen strip chart paper recorder.  The radar also had a needle meter calibrated in mph.  Some systems had 2 antennas, one for transmitting and the other for receiving.  The antennas were mounted on a tripod, patrol vehicle hood, or fender.  Some of the early 1960s' models mounted the antennas in the back windshield of the patrol car.

The first traffic radars transmitted at 2.455 GHz in the S band (2 - 4 GHz).  The antenna beamwidth varied from 15 to 20 degrees depending on model.  These radars operated from a stationary position only and measured receding as well as approaching traffic to an accuracy of about ± 2 mph.  The maximum detection range was an unimpressive 150 to 500 feet, vacuum-tube receivers do not have the sensitivity of solid-state receivers.  S band radars are obsolete.

## X Band Radar
### 10.525 GHz ± 25 MHz

X band radars have been around since 1965 and operate on a single frequency, one 50 MHz channel.  Radars in the X band have better all weather performance, less signal attenuation in bad weather than higher frequency systems in K, Ka, and infrared bands.  The X band radars tend to have slightly wider beams.

Some European countries use X band traffic radars that transmit at **9.41 GHz** or **9.90 GHz**.

## Ku Band Radar
## 13.45 GHz

The Federal Communications Commission has allocated 13.45 GHz in the Ku band for police radar use in the United States, however Ku radars are not sold or used in the U.S. Some European and Middle Eastern countries use Ku band traffic radars.

## K Band Radar
### 24.150 GHz ± 100 MHz
### 24.125 GHz ± 100 MHz

K band radars have been around since 1976 and operate on a single frequency, one 200 MHz channel.

These radars generally have more narrow beams than X band radars, and slightly wider beams than Ka band radars.

Detection range decreases with moisture. The water vapor absorption band is centered at 22.24 GHz, signals in this band tend to become absorbed by moisture in the atmosphere. For short range applications the effects may be tolerable on relatively clear dry days.

# Ka Band Radar
# 33.4 - 36.0 GHz

In 1983 the United States Federal Communications Commission allocated the spectrum from 34.2 - 35.2 GHz for police radar use. That same year Ka band photo *across the road radars* started appearing in the United States. Nine years later in 1992 the FCC expanded the Ka band spectrum allocated for police radar to 33.4 - 36 GHz.

Ka band radars typically have more narrow beams than X or K band radars. Detection range depends on moisture in the atmosphere, the more moisture the shorter the detection range.

Many models have a channel bandwidth of ±100 MHz, a 200 MHz wide channel. Some models channel band is ±50 MHz, a 100 MHz wide channel. This gives Ka band radar multiple channels, 13 to 26 channels depending on channel bandwidth. Most Ka band police radars operate on one frequency channel, a few have 2 channels an operator can select.

Wideband Ka Radar
Wideband Ka radars operate on a single fixed frequency, or hops between one or more other frequencies. In frequency hop mode the radar dwells on one frequency for one or more sample periods then switches to another. The frequency hopping mode is intended to defeat radar detectors, however is seldom if ever used due to problems to numerous to mention. Only a few police radar models have this mode.

## Photo Radar

Photo radars are **Across the Road Radars** that intentionally aim a narrow beam across the road instead of *down the road*. The main beam of the radar radiates only a very small segment of the road. These systems account for the cosine effect angle and adjust measured speed upwards 6% - 9%.

Across the road radars are **inherently less accurate** than down the road radars because the beam is angled to traffic direction. The angle causes the Doppler reflection to spread out as a vehicle passes through the beam, a built-in speed error. At speeds as low as 20 mph the speed spread is 6 mph, the faster the speed the greater the spread. The radar attempts to process out the spread, but there still exists more uncertainty compared to down the road radars.

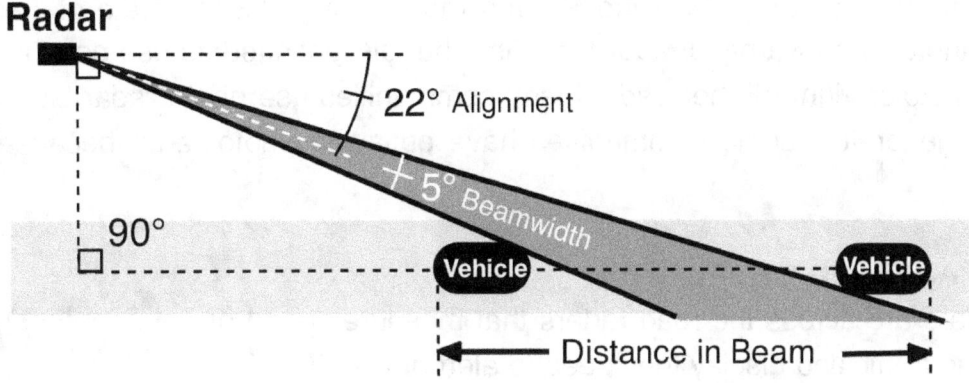

Figure 1.1-2 -- **Across the Road Radar**

Photo radars can be mounted in patrol vehicles but can only be used from a stationary position. Some are designed to be mounted to poles, portable tripods, or inside ground fixtures sometimes disguised as trash containers, billboards or other common objects. Photo radars have also been hidden in highway maintenance and construction vehicles, tractor-trailers, box trucks, and unmarked passenger vehicles including vans.

Photo radars must be properly aligned to the road for the radar to correctly process the cosine effect. If the alignment angle is shallow measured speed is high, if the angle is larger measured speed is low.

Most if not all fixed mounted photo radars operate unattended, no police officer or operator present. Mobile photo radars can operate either unattended or attended. In unattended operation the radar is constantly transmitting. When an operator is present the radar can run all the time, or the operator can use the instant-on feature to transmit only on command.

Violations are photographed and superimposed with speed, date, time and location and mailed to vehicle *owner.* Photo radar setup to measure approaching traffic gets the front license plate, the driver may or may not be identifiable. Some systems take a second photograph to get the rear license plate. Photo radar setup to measure receding traffic only gets the rear license plate.

Photo radars, or camera radars as they were first called, were in experimental stages of development as early as 1954 using S band radars. In 1983 the state of Texas tried a French-manufactured Ka band radar for a time but discontinued its use because the units were being stolen right off the road. Many communities use photo radar because of the revenue it generates, some communities have outlawed photo radar because of public pressure.

## SAFETY RADAR

Safety radars are across the road radars that measure speed of approaching traffic and display the speed to alert drivers how fast they are traveling. These radars are constantly transmitting. Most safety radars display all speeds measured, some only display speeds above a preset limit. Many safety radars do not record anything, some record all speeds measured and some only record speeds above a preset limit.

Safety radars can be built into small portable trailers, larger trailers with larger displays, made in the shape of various road signs and other objects. In the United States most operate in the K or Ka band because the X band is overcrowded.

## TRAFFIC STUDY RADAR

Some across the road radars do not display speed but do record speed, date, time, and location for traffic studies. The radars are constantly transmitting, and measure approaching and/or receding traffic. The purpose of traffic study radar is not to issue tickets or show speed, only to estimate traffic speeds on a given road at given times.

# Laser Radar

Laser radars or lidars were introduced in the early 1990s. These systems radiate in the upper infrared band and have extremely narrow beams compared to microwaves. Lidars transmit a narrow infrared laser pulse to measure pulse round trip time to a target to calculate target range. Speed is calculated from the change in range with time.

Laser radars are hand held and function from a stationary position only, no moving mode, and measure speed and range of approaching or receding traffic. Laser radars can also measure the range of stationary objects.

Lidars have shorter detection range and are much more sensitive to weather conditions than microwave radars. Laser signals propagate best in clear dry cool atmospheric conditions.

# Microwave versus Laser Radar

Police microwave radars do not require the operator to aim exactly at a particular vehicle, only the general direction plus or minus half a beamwidth (± 4° to ±10°).  These radars are most effective in light to moderate traffic at short and long ranges.  Many microwave radars can operate from a moving patrol vehicle.

Police laser radar must be aimed exactly using crosshairs or aim dot at at flat surface on a particular vehicle.  These radars are most effective at short ranges in light or dense traffic.  Laser radars are not designed to operate from a moving vehicle.

|  | MICROWAVE RADAR | LASER RADAR |
|---|---|---|
| **Operation** | Stationary or Moving | Stationary only |
| **Aim** | Easy Aim | Exact Aim Required |
| **Transmissions** | Continuous or Instant-On | Instant-On only |
| **Traffic Conditions** | Light - Moderate Traffic | Light or Dense Traffic |
| **Range** | Short or Long Range Traffic | Short Range Traffic |
| **Measurements** | Speed only | Speed and Range |
| **Location** | Inside or Outside Patrol Vehicle | Outside Patrol Vehicle<br>Should not be operated from behind glass / windshields etc. |

Table 1.1-1 -- **Microwave versus Laser Radar**

# Chapter 1.2 -- Timing Systems (Clocking)

A number of speed measuring systems measure the time it takes a vehicle to travel a known distance to calculate speed. Distance traveled equals speed multiplied by time, speed equals distance divided by time.

$$Distance = Speed \times Time$$
$$Speed = Distance / Time$$
$$\mathbf{v = d / t}$$

v = Vehicle Speed          t = Time
d = Distance Traveled

All timing and distance measuring instruments should be calibrated for accuracy on a periodic basis. Testing data should include at the least; dates tested, test facility, test methodology and data, and due date for next test.

# Automatic Timing Systems

## Inductive Loops

An inductive loop uses 2 sensors spaced typically ten's of feet apart. Each sensor is a conductive wire, cable, pipe or metal strip in the shape of a relatively long narrow rectangular loop buried in and across the road. Some portable systems consist of two rubber type strips placed on top of and perpendicular to the road. Each sensor is electrically charged and generates a magnetic field directly above the loop. When a vehicle enters the magnetic field the vehicle changes the magnetic field which is detected by the sensor. The time it takes a vehicle to pass each sensor is used to calculate speed.

## Magneto Sensors

Magneto resistive sensor systems use 2 sensors spaced typically ten's of feet apart buried into the road. The time it takes a vehicle to pass each sensor is used to calculate speed. Each sensor is a cylindrical shape with a 4 inch diameter and 2 inches deep. The sensors are battery powered and communicate to a processor using a wireless link.

## Piezo Ceramic Sensors

Piezo ceramic sensor systems use 2 sensors spaced typically ten's of feet apart buried in or flush with the road. Each sensor covers an entire lane and is roughly several inches wide. The sensor detects pressure from a vehicle's weight. The time it takes a vehicle to pass each sensor is used to calculate speed. The sensor only measures the vehicle front tires because the ceramic sensor cannot recover to a steady state, zero pressure, fast enough to measure rear tires.

## Road Cables

Road cables are portable systems that consist of 2 or more pneumatic cables placed across the road typically ten's of feet apart. The time it takes a vehicle to pass each cable is used to calculate speed. *Each pneumatic cable MUST be the same length*, different lengths result in different trigger times introducing an error. Shorter cables trigger faster than longer cables.

## Across the Road Laser

Across the road laser uses two independently pulsed laser beams that cross the road at a 90 degree angle. The system measures the time it takes the front of a vehicle to travel from the first beam to the second beam. Additionally a second measurement tracks when the rear of the vehicle clears both beams. The two measurements must match closely or the data is rejected. Two measurements are taken in the event multiple vehicles cross the beams.

# Manual Timing Systems

## Aerial Clocking

Aerial Clocking is probability the best system for police in terms of catching large numbers of speeders, but one of the worst in terms of personnel, equipment and expense. A police officer observes traffic from an airplane or helicopter and times targets as they pass between two points, typically lines painted across the road spaced 1/10 mile apart. The airborne officer times targets using a timing computer or a stop-watch. The officer radios to patrols on the ground which vehicle to ticket when a violation is calculated.

## Distance Time Computer

Distance / Time computers can be hand held or built into a patrol vehicle such as VASCAR. Some microwave and laser radars have an optional timing mode to measure speed without transmitting. The operator inputs distance between points and uses the computer to time targets by pushing a button when a target passes each point. Most lidars have a range only mode to allow the operator to measure distance from lidar to an object for use in manual timing mode.

## VASCAR

VASCAR (**V**isual **A**verage **S**peed **C**omputer **A**nd **R**ecorder) is a distance / time computer built-in to the patrol vehicle. Distance is manually entered into the computer, or by measuring distance by driving pass two points. A button is pushed at each point so distance can be derived from patrol vehicle odometer. The operator inputs the time parameter by pushing another button when the target passes each point.

## Stop Watch

A stop-watch can be used to calculate speed by measuring the time it takes a target to travel between 2 points of known distance. The greater the distance the more accurate the measurement.

$$\text{Speed in MPH} = (15/22) \times \text{Distance} / \text{Time}$$
Distance in **feet**, time in **seconds**.

$$\text{Speed in KPH} = 3.6 \times \text{Distance} / \text{Time}$$
Distance in **meters**, time in **seconds**.

# Timing System Errors

Timing systems are vulnerable to any errors in measured time or distance.  Timing vehicles over longer distances and longer times tends to reduce errors.  Plus or minus a fraction of a second error over a long distance introduces a smaller error than for a short distance.

- If measured time is short, calculated speed is faster.
- If measured time is long, calculated speed is slower.

- If distance used is short, calculated speed is slower.
- If distance is long, calculated speed is faster.

Errors in both time and distance can add or subtract from actual speed.  If time short and distance long, or time long and distance short, the combination error will partially or fully cancel out.

The faster a target the greater the time and distance required to calculate an accurate speed measurement.  For most cases a small distance error produces a small speed error, however small time errors can produce significant speed errors in many situations.

## Measured Speed

$$v_m = (d_o \pm d_{err}) / (t_o \pm t_{err})$$

$$\mathbf{v_m = (d_o \pm d_{err}) / (d_o/v_o \pm t_{err})}$$

| | |
|---|---|
| $v_m$ = Measured Speed | $d_o$ = Distance traveled in Time $t_o$ |
| $v_o$ = Actual Speed | $t_o$ = Time vehicle travels Distance $d_o$ |
| | $t_{err}$ = Time Error |
| | $d_{err}$ = Distance Speed |

---

Measured Speed in MPH = $15 (d_o \pm d_{err}) / (15 d_o/v_o \pm 22 t_{err})$

Distances in **feet**, times in **seconds**, speeds in **MPH**

---

Measured Speed in KPH = $3.6 (d_o \pm d_{err}) / (3.6 d_o/v_o \pm t_{err})$

Distances in **meters**, times in **seconds**, speeds in **KPH**

# Chapter 1.3 -- Pacing

Pacing with the speedometer involves the patrol car following a vehicle while maintaining a constant range separation. This method depends on the ability of the officer to maintain patrol car speed with target speed and the accuracy of the patrol car speedometer. Some agencies recommend a target vehicle should be paced for at least 0.2 miles, and only forward vehicles and not vehicles in the rear view mirror.

## Speedometers

There are 2 basic types of speedometers, mechanical and electronic. The oldest type of speedometer is mechanical and still used. Electronic speedometers, also called Vehicle Speed Sensors (VSS), started showing up in the 1980's and are common on most new vehicles.

MECHANICAL -- Mechanical speedometer's use a *flexible rotating cable* usually connected by gears or sometimes magnets to the transmission case vehicle drive shaft or differential assembly. Many motorcycles and early Volkswagen Beetles connect to a front wheel.

ELECTRONIC -- Vehicle Speed Sensors use *electrical signals* from sensors, usually magnetic but sometimes gears, mounted to the transmission case vehicle drive shaft or differential assembly. Some VSS's use *wheel sensor data* to calculate speed, some use Anti-lock Brake System (ABS) data. Some police radar's use VSS data to detect patrol speed shadowing in moving mode operation.

All speedometers should be calibrated on a periodic basis, and repaired if out of factory specifications. Specified accuracy varies with make and model and can easily be ± 2 mph or more. There should be a calibration paperwork trail that includes at the very least;

- Date Tested,
- Testing facility,
- Test Method - dynamometer or road test with speed wheel,
- Test Data,
- Next Test Due Date.

## Speedometer Accuracy

Tire pressure and diameter varies with use and conditions.  As long as the tires are maintained with proper cold tire pressure and used within specifications, not exceeding maximum load and speed, speedometer accuracy will be within factory specifications.  The greater the actual diameter is different from the standard speedometer calibrated diameter, the greater the measured speed error.

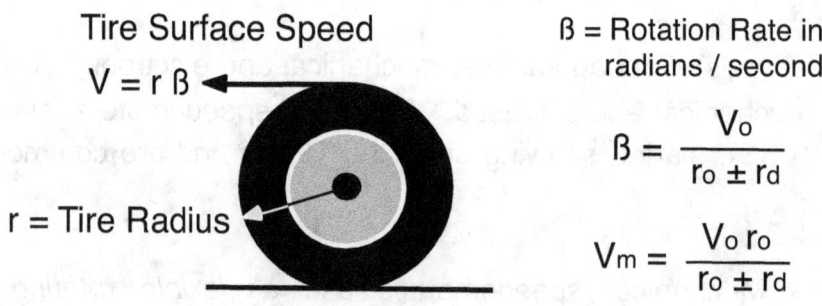

Tire Surface Speed

$$V = r \beta$$

r = Tire Radius

$\beta$ = Rotation Rate in radians / second

$$\beta = \frac{V_o}{r_o \pm r_d}$$

$$V_m = \frac{V_o r_o}{r_o \pm r_d}$$

Figure 1.3-1 -- **Tire Radius and Measured Speed**

$V_m$ = speedometer measured speed          $r_o$ = tire specified radius
$V_o$ = true speed                          $\pm r_d$ = tire radius difference

### Measures Speed based on Tire Diameter

$$V_m = V_o \left( d_o / d_a \right)$$

$$V_{err} = 100 \left[ \left( d_o / d_a \right) - 1 \right) \right] \%$$

$d_o$ = specified tire diameter          $d_a$ = actual tire diameter

$V_m$ in same units as $V_o$ (mph, kph, etc.), $d_o$ and $d_1$ must be in same units (inches, feet, millimeters, meters, etc.).

## Tire Specifications

Many tires have sidewall markings that list tire width, aspect ratio or sidewall height to width, and wheel diameter. Tire diameter can be derived from width, aspect ratio, and wheel diameter.

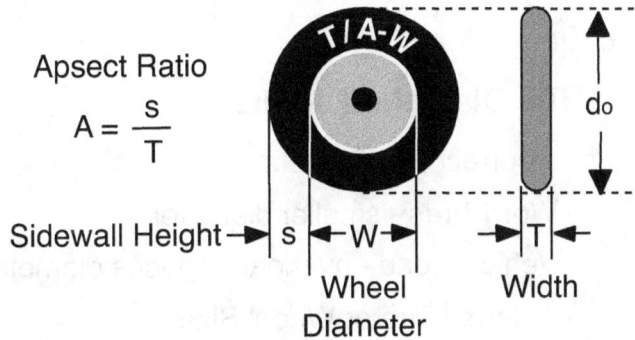

$$A = \frac{s}{T}$$

### Tire Diameter ($d_o$)

$$d_o = 2\,s + W$$
$$= 2\,T\,A + W$$

The International Organization for Standardization (ISO) uses a T/A_W format, usually embedded in the tire code.

### Tire code = P 185 / 60 R 15 84 T

### T / A _ W = 185 / 60 _ 15

| T | 3 digit number | Nominal Width at widest point in millimeters (mm). |
|---|---|---|
| A | 2 or 3 digit number | Aspect ratio (sidewall height / width) as a percent. If omitted = 82%, if > 200 diameter in mm. |
| W | 2 digit number | Wheel Diameter in inches. |

Table 1.3-1 -- **Tire Size ISO Code**

### Tire Diameter

$d_o$ in inches = **(TA/1270) + W**

$d_o$ in mm = **(0.02 TA) + (25.4 W)**

## Tire Factors

All speedometers depend on the tires maintaining a constant diameter.  If the actual tire diameter is smaller than the standard tire diameter, speed measures HIGH.  If actual diameter is greater than standard, speed measures low. The greater the difference the greater the speed error.

### Tire Diameter Factors

- Incorrect Tire Pressure
- Worn Tires - smaller diameter
- Vehicle Load - overload reduces diameter
- Different Tire or Wheel Size
- Altitude
- Tire Temperature
- Atmospheric Pressure
- Different Differential Gearing

# Chapter 2 -- The Cosine Effect

## Chapter 2.1 -- Cosine Effect Error

Microwave and Laser Radar

## Cosine Effect Setup

Police microwave and laser radars measure the relative speed a vehicle is approaching, or receding, the radar. If a vehicle is traveling *directly at* the radar the relative speed is actual speed. If the vehicle is not traveling directly at the radar the relative speed is slightly lower than actual speed. The phenomenon is called the **Cosine Effect** because the measured speed is directly related to the cosine of the angle between the radar and vehicle direction of travel or speed vector. The greater the angle the greater the speed error and the lower the measured speed. A cosine angle of 90° has 100% error, speed measures zero.

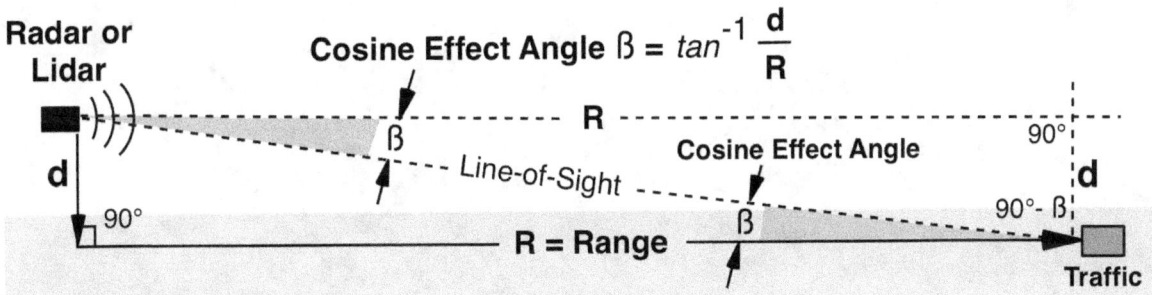

$$\text{Measures Speed} = V_m = v_o \cos \beta = v_o \frac{R}{\left(R^2 + d^2\right)^{0.5}}$$

Figure 2.1-1 -- **Cosine Effect Setup**

$v_o$ = Speed                        R = Range
$\beta$ = Cosine Effect Angle        d = Antenna Distance to Traffic Lane Center

The cosine effect angle is the angle between the radar antenna and the vehicle direction of travel.  The angle depends on vehicle range and radar antenna distance from vehicle lane center.   The antenna pointing angle is completely irrelevant, only the angle to the target vehicle matters.

The cosine function is always between 0 and 1.  The cosine of 0° is 1, measured speed is actual speed.  The cosine of 90° is 0, measured speed is zero.  The closer the angle is to 0° the more accurate the speed measured.

The cosine effect applies to both microwave and laser radars.  Photo radars are *Across the Road Radars* that angle the beam across the road, typically 20° to 23°.  As a vehicle passes through the beam the measured speed is changing and spreading due to the *cosine effect*. The changing and spreading of the measured speed in a very short time makes photo radars inherently less accurate than *Down the Road* microwave and laser radars.

## Overpass

The cosine effect has 2 components if the radar or lidar is elevated to traffic such as from an overpass. Radar distance from traffic lane is a function of horizontal and vertical distances.

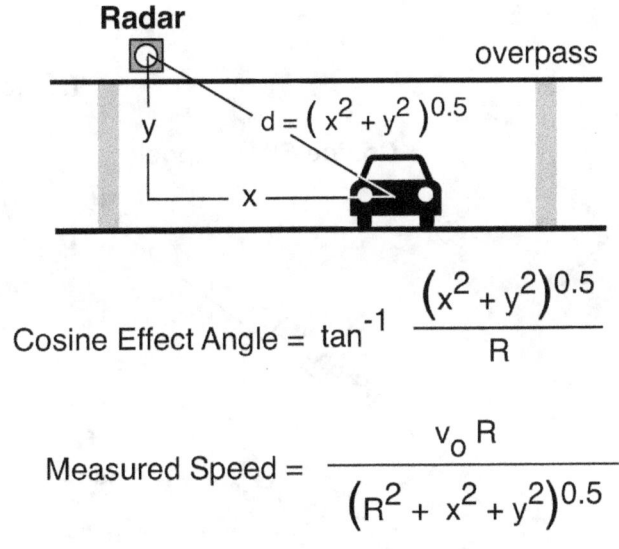

**Radar**

overpass

$$d = \left( x^2 + y^2 \right)^{0.5}$$

$$\text{Cosine Effect Angle} = \tan^{-1} \frac{\left(x^2 + y^2\right)^{0.5}}{R}$$

$$\text{Measured Speed} = \frac{v_o R}{\left(R^2 + x^2 + y^2\right)^{0.5}}$$

Figure 2.1-2 -- **Cosine Effect from an Overpass**

d = radar distance to vehicle lane
x = radar horizontal distance to lane center
y = vertical distance to height of vehicle

R = vehicle range to radar
$v_o$ = vehicle speed

In the figure, "x" is the horizontal distance to the vehicle lane center, and "y" is the vertical distance to the average top of the target vehicle. Radar distance to vehicle path is "d" and equals the square root of the sum of "x" squared plus "y" squared. If either the horizontal or vertical component is zero, the equations reduce to that shown in Figure 2.1-1.

## Traffic on a Curve

The cosine effect angle on curves is large and changing proportional to curve radius and vehicle speed. The relatively large and fast changing cosine effect angle results in measured speed changing relatively fast, too fast in most scenarios for a radar or lidar to measure speed.

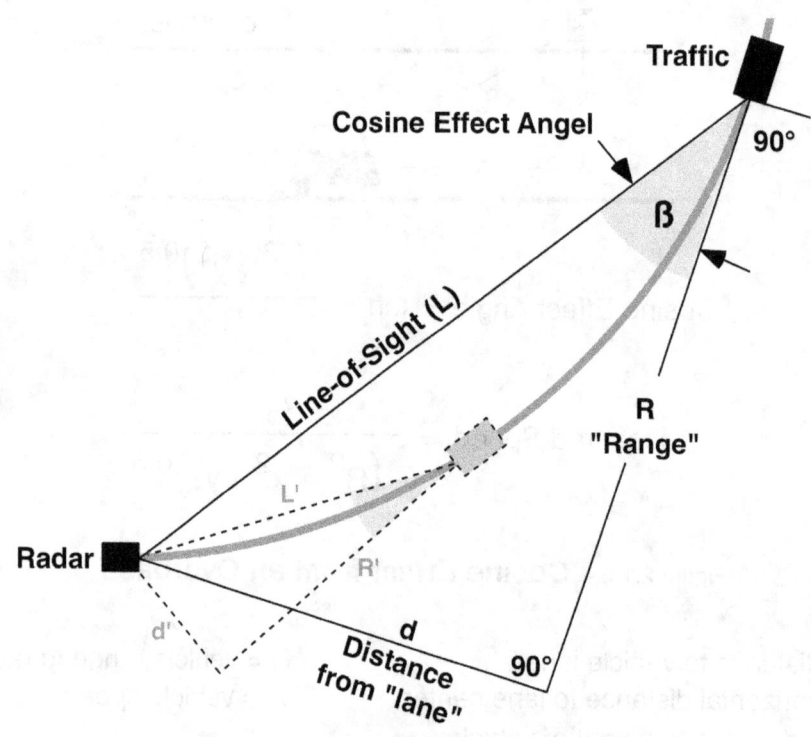

Figure 2.1-3 -- **Changing Cosine Effect Angle**

$$ß = \tan^{-1}(d / R)$$

$$ß' = \tan^{-1}(d' / R')$$

Microwave and laser radars cannot measure traffic speed on a curve because the angle is changing causing the relative speed to change too fast for the radar or lidar to measure.

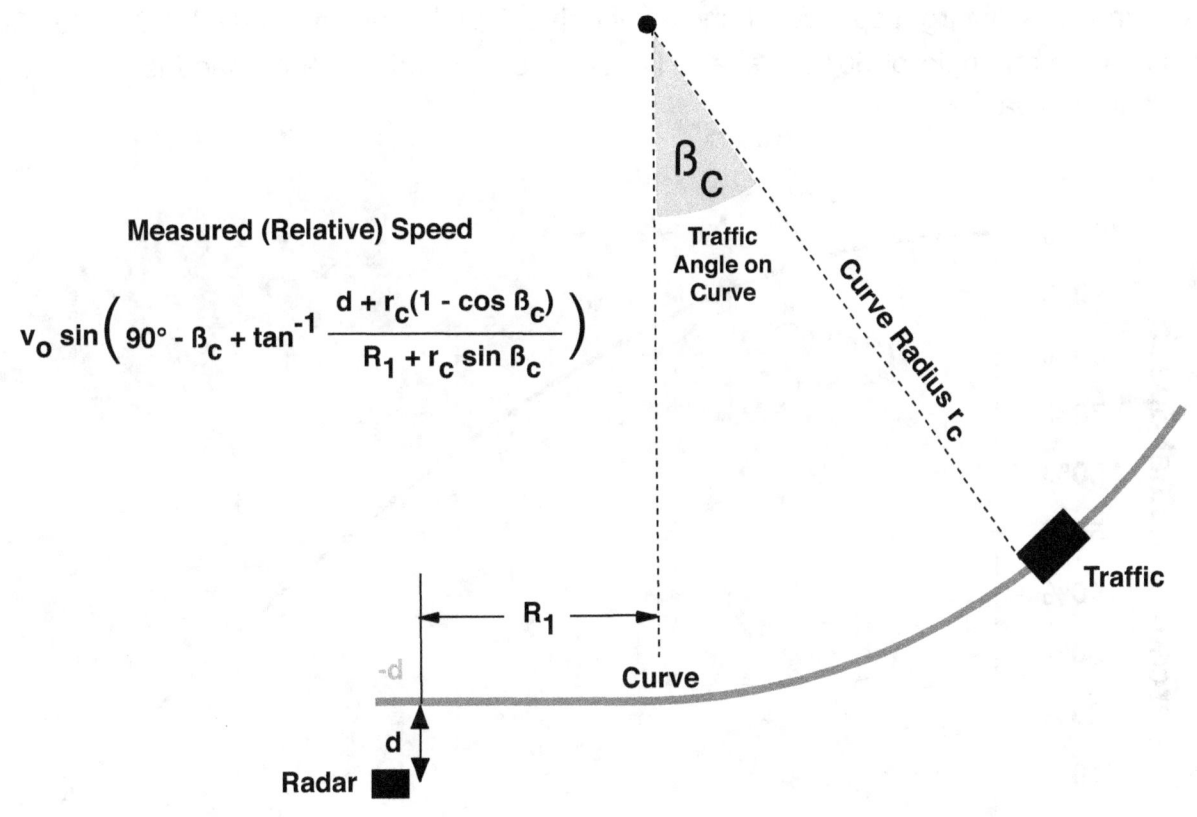

**Measured (Relative) Speed**

$$v_o \sin\left(90° - \beta_c + \tan^{-1} \frac{d + r_c(1 - \cos \beta_c)}{R_1 + r_c \sin \beta_c}\right)$$

Figure 2.1-4 -- **Measured Speed on a Curve**

**Measured Speed Variables**

- Vehicle Speed ($v_o$)
- Vehicle Angle on Curve ($\beta_c$)
- Curve Radius ($r_c$)
- Radar Range to start/end of Curve ($R_1$)
- Radar Distance to Vehicle Lane (d)

## Cosine Error Graph

The below figure is a graphical representation of the Cosine Effect for measured speed as a percentage of true speed versus angle.  The larger the angle the larger the error and the lower the relative target speed.  At angles of only a few degrees the speed is 99% to 100% of actual, at an angle of 60° the speed is half actual speed.  At 90° speed is 0 relative to the radar or lidar.

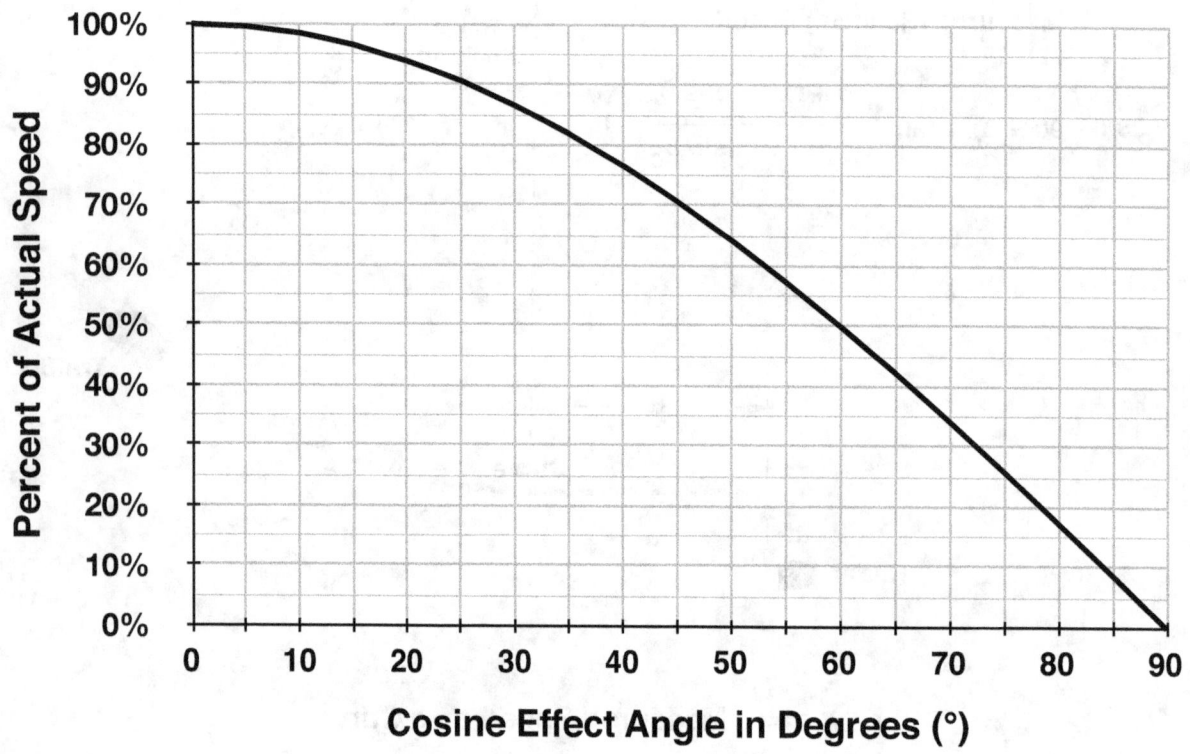

Figure 2.1-5 -- **Cosine Effect Angle vs Measured Speed Percent**

## Chapter 2.2 -- Cosine Effect Minimum Range

Microwave and Laser Radar

## Cosine Effect Acceleration

The cosine effect induces a deceleration component (-a) for approaching vehicles, and an acceleration component (+a) for receding vehicles.  The closer the vehicle the faster the angle and measured speed change and the greater the acceleration component.

The cosine effect acceleration component is a function of vehicle speed and range from radar, and radar distance from vehicle lane.  Applying a little calculus and trigonometry to velocity vectors and the geometry the vehicle cosine effect acceleration can be calculated.

**Acceleration based on target vehicle speed and range from radar, and radar distance from target vehicle lane.**

$$a = \frac{v_o^2 \, d^2}{\left( R^2 + d^2 \right)^{1.5}}$$

a = acceleration
d = radar distance from target vehicle lane center
$v_p$ = target vehicle speed
R = range

The cosine effect equation for acceleration can be used to determine the range for a given acceleration and radar distance from vehicle lane and vehicle speed.

**Range based on target vehicle speed, radar distance from target vehicle lane, and acceleration.**

$$R = \left( \frac{(v_o \, d)^{4/3}}{a^{2/3}} - d^2 \right)^{0.5}$$

R = vehicle range from radar
$v_p$ = vehicle speed
d = radar distance from target vehicle lane center
a = cosine effect acceleration

The above equations can be used to determine radar minimum range, vehicles closer than minimum range cannot be measured.


## Minimum Range

A radar cannot measure speed when acceleration exceeds the radar limits due to the cosine effect, or anything else - curves, vehicle slowing or accelerating. Radar acceleration limit is a function of radar accuracy and sample time - minimum time required to get one speed measurement. Speed cannot change faster than radar or lidar accuracy during one sample period.

**Radar / Lidar Acceleration Limit**

$$a_{max} = \pm \, v_{acc} \, / \, t_i$$

$a_{max}$ = acceleration limit
$v_{acc}$ = accuracy
$t_i$ = sample time or integration period

The acceleration limit sets a minimum range, vehicles inside minimum range are too close to measure because the relative speed is changing too fast.  Minimum range applies to vehicles traveling toward and away from the radar.

**Minimum Range Factors**

• **Traffic Speed**

• **Antenna Distance from Traffic Lane**

• Radar Accuracy (typically ± 1 mph)

• Radar Sample Time (typically 0.3 seconds)

Accuracy is typically ±1 mph, sample times vary with model from 1/4 to 1/2 second. Many microwave radars have a sample time of 0.3 seconds (300 milliseconds), laser radars typically have longer sample periods.

Target vehicle acceleration must be within limits for an entire sample time. Radar / lidar minimum range is the range acceleration exceeds limits plus one sample period.

## Minimum Range Equation

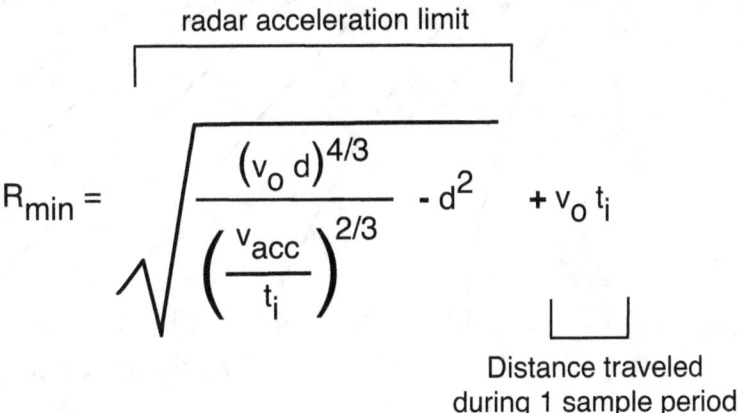

radar acceleration limit

$$R_{min} = \sqrt{\frac{(v_o\, d)^{4/3}}{\left(\dfrac{v_{acc}}{t_i}\right)^{2/3}} - d^2} \; + v_o\, t_i$$

Distance traveled
during 1 sample period

$R_{min}$ = Minimum Range        $v_{acc}$ = Radar Accuracy
$v_o$ = Speed                 d = Radar Distance from Target Vehicle Lane Center
                         $t_i$ = Radar Sample Time

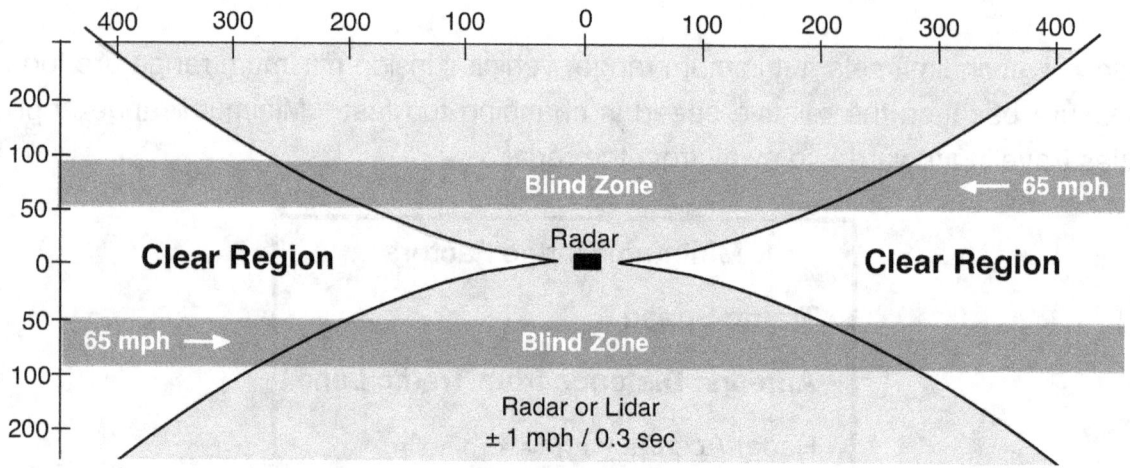

Figure 2.2-1 -- **MInimum Range Example**
Distances in feet

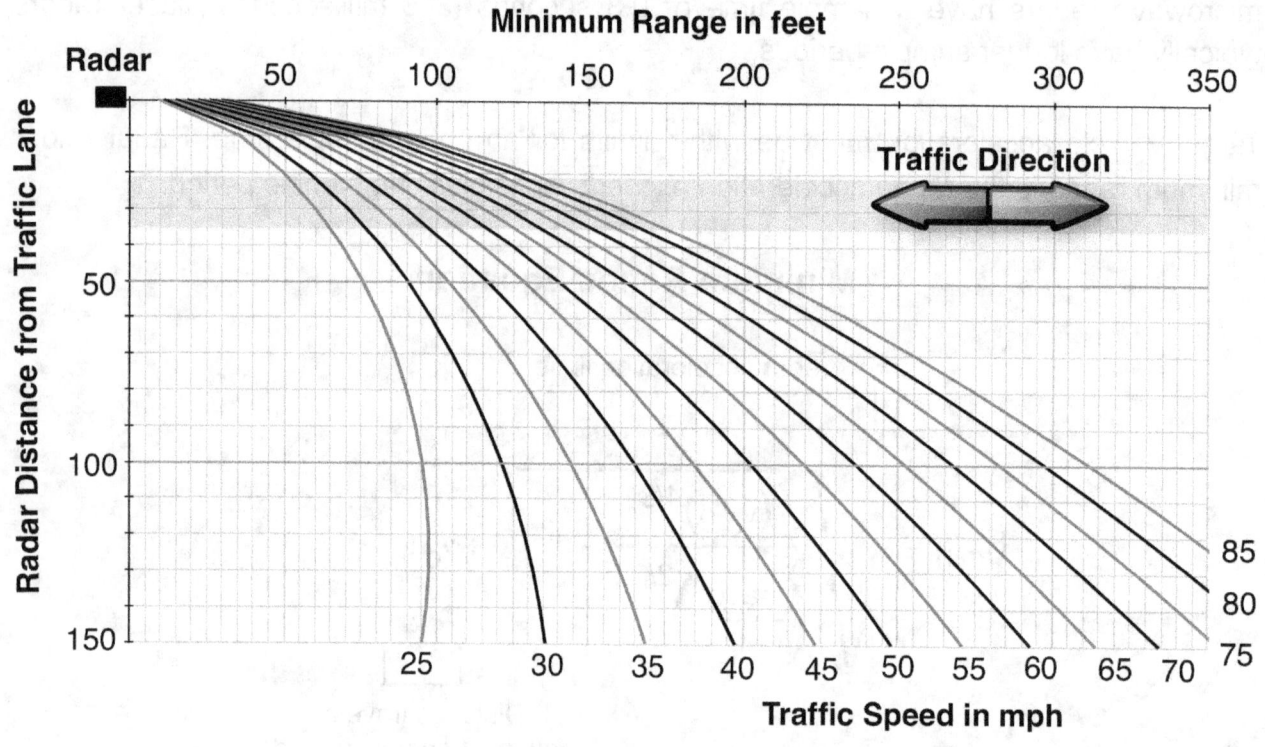

Figure 2.2-2 -- **MInimum Range**
**0.250 second Sample Time and ± 1 mph Accuracy**

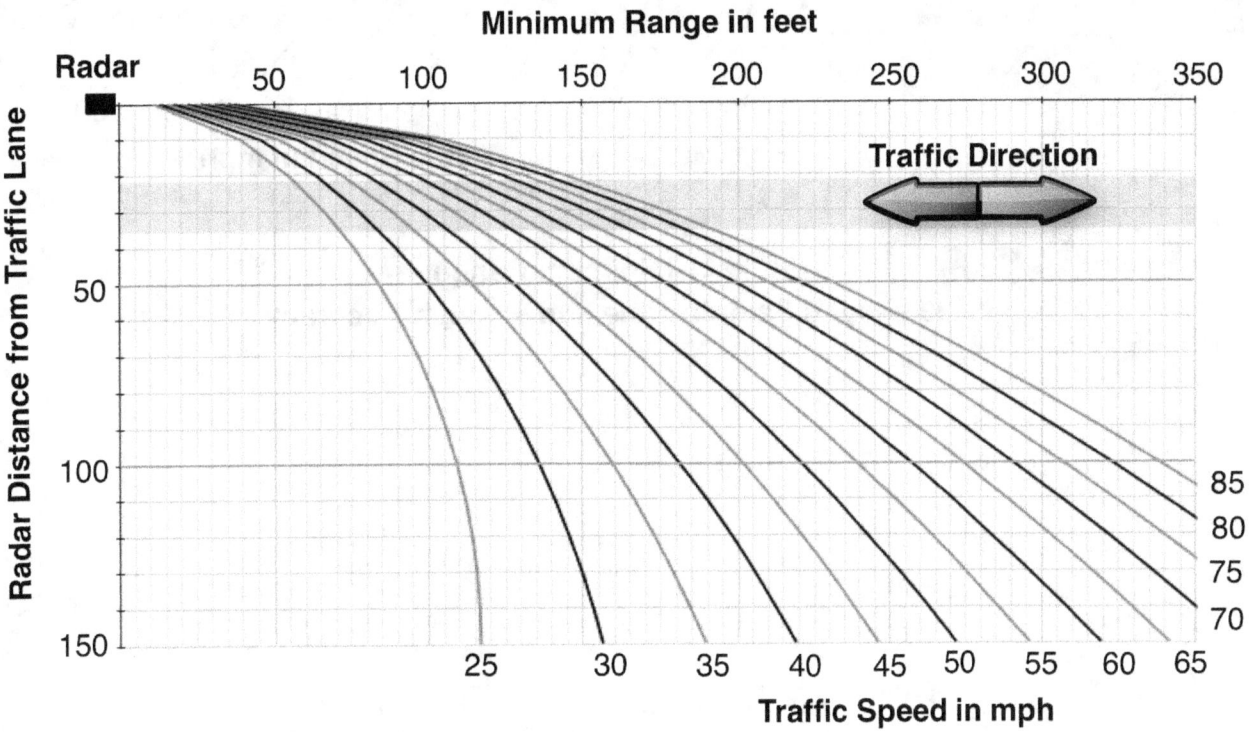

Figure 2.2-3 -- **MInimum Range**
**0.300 second Sample Time and ± 1 mph Accuracy**

Many microwave radars use a 0.3 second sample time.

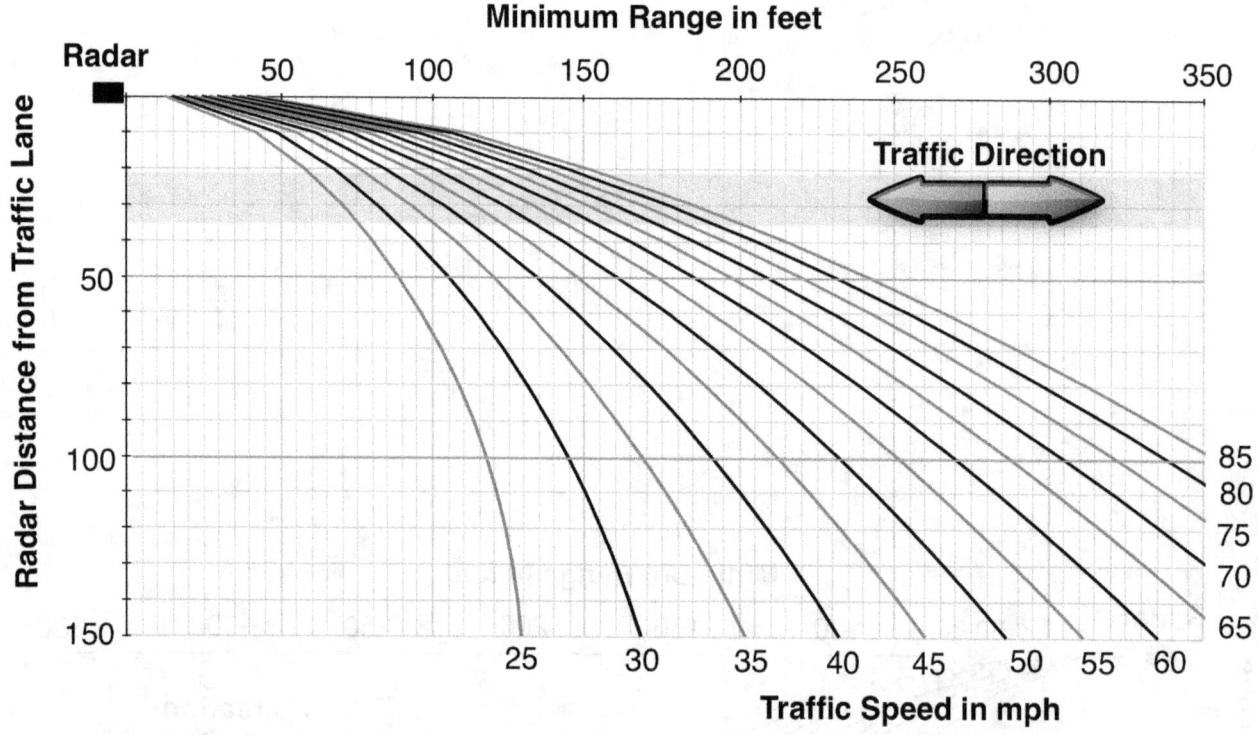

Figure 2.2-4 -- **MInimum Range**
**0.333 second Sample Time and ± 1 mph Accuracy**

# Chapter 2.3 -- Moving Radar Cosine Effect

Microwave Radar

## Cosine Effect Error

The main difference between stationary and moving mode radar is moving radar also measures patrol vehicle speed by measuring the ground reflection speed. A moving radar measures traffic speed relative to the radar, closing speed for on-coming traffic. Patrol vehicle speed is subtracted from closing speed to obtain traffic speed.

| **Closing Speed** | **Target Vehicle Speed** |
|:---:|:---:|
| $V_c = V_o + V_p$ | $V_o = V_c - V_p$ |

$V_c$ = closing speed    $V_p$ = patrol speed,    $V_o$ = target vehicle speed

Moving mode radar antennas are designed to be aimed exactly in the direction of travel, directly forward for front antennas and directly aft for rear antennas. An off angle ground echo measures patrol speed low by the product of the cosine angle resulting in a traffic speed error, *Patrol Speed Shadowing* by definition.

### Radar Measured Patrol Speed

$$V_{pm} = V_p \, cos \, ß_p$$

$V_{pm}$ = Radar Measured Patrol Speed          $ß_p$ = Ground Echo Angle Error
$V_p$ = Actual Radar Patrol Speed

**A patrol vehicle speed that measures low will calculate target vehicle speed high by the speed the patrol vehicle is low.**

The radar identifies the main beam center ground echo as the strongest echo, it *usually* is by several orders of magnitude. Sometimes stronger reflections come from off angle that may be at the main beam edge instead of center, or may not be in the main beam at all but in a side lobe.

## Opposite Direction Traffic

The cosine effect potentially introduces 2 errors in moving mode. One error is the angle between antenna and target vehicle which measures target speed low. The other error is the angle of the ground echo, the greater the angle the lower measured patrol speed and the higher the measured target speed.

### Moving Mode Speed Angle Errors

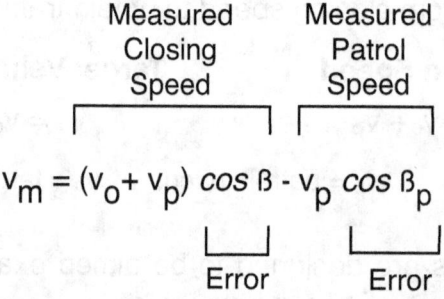

$$v_m = (v_o + v_p)\ cos\ \beta - v_p\ cos\ \beta_p$$

$v_m$ = measured traffic speed          $\beta$ = cosine effect angle

$v_o$ = traffic speed          $\beta_p$ = ground echo angle error

$v_p$ = patrol vehicle speed

### Measured Speed Error

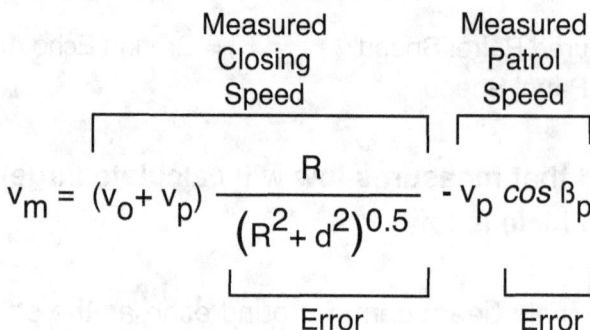

$$v_m = (v_o + v_p)\ \frac{R}{(R^2 + d^2)^{0.5}} - v_p\ cos\ \beta_p$$

$v_m$ = measured traffic speed          $R$ = vehicle range from radar

$v_o$ = traffic speed          $d$ = radar distance from vehicle lane

$v_p$ = patrol vehicle speed          $\beta_p$ = ground echo angle error

## Cosine Effect Acceleration

The cosine effect on acceleration for moving mode is the same as for stationary operation except the patrol vehicle radar and traffic are approaching, or receding, each other. The cosine effect angle is changing at the rate proportional to not just traffic speed but also the patrol vehicle speed. The acceleration equation for stationary radar applies to moving mode radar by substituting closing or opening speed for traffic speed.

$$a = \frac{(v_o + v_p)^2 \, d^2}{\left( R^2 + d^2 \right)^{1.5}}$$

a = cosine effect acceleration      R = vehicle range from radar
$v_o$ = traffic speed      d = radar distance from vehicle lane
$v_p$ = patrol vehicle speed

## Cosine Effect Minimum Range

Moving mode radar acceleration limit is a function of radar *patrol speed accuracy* and minimum *sample time*. Typically traffic accuracy is ±2 mph and patrol accuracy is ±1 mph. The radar's traffic speed measurement accuracy for moving mode is double that of stationary mode because the radar is using 2 measurements, radar speed and closing or oening speed to derive traffic speed - 2 error sources.

### Radar Acceleration Limit

$$a_{max} = \pm \, v_{acc} / t_i$$

$a_{max}$ = acceleration limit
$v_{acc}$ = patrol speed accuracy
$t_i$ = sample time

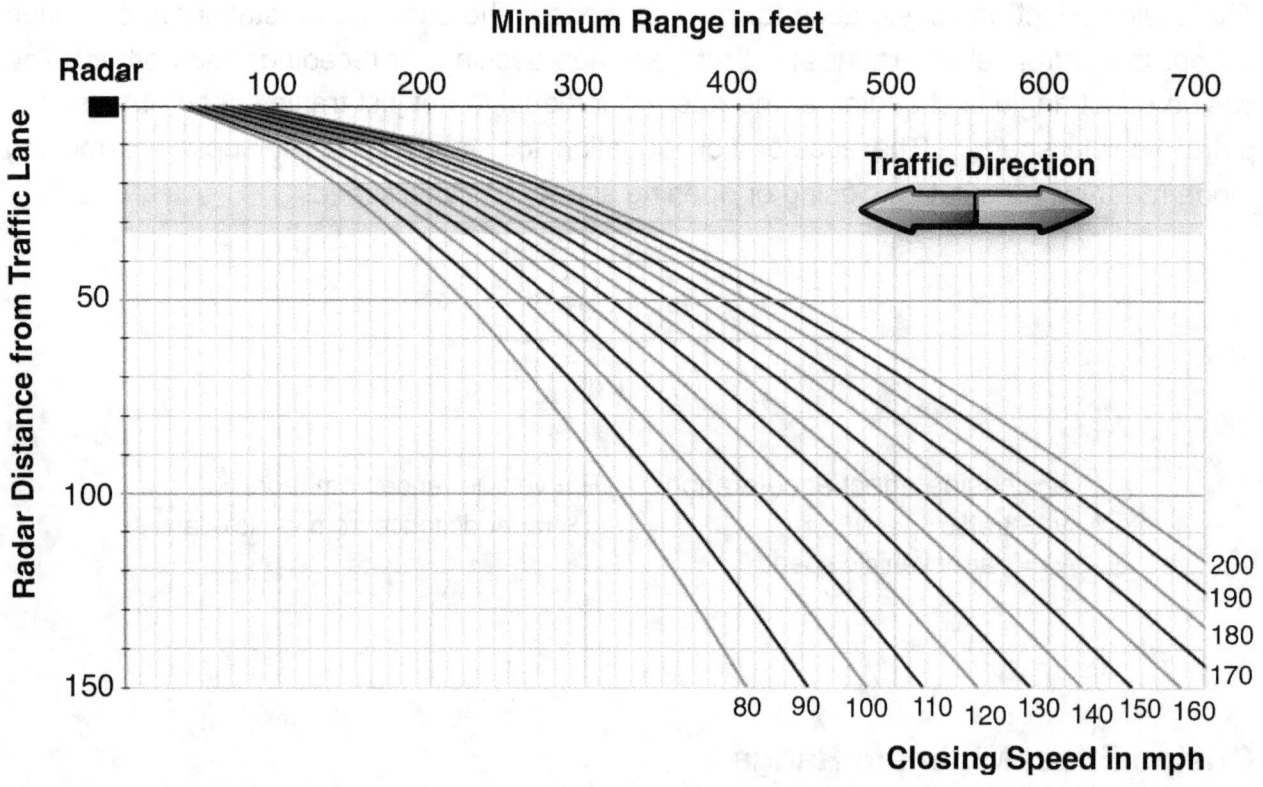

Figure 2.3-1 -- **Minimum Range**
**0.300 second Sample Time and ± 1 mph Accuracy**

Many microwave radars use a 0.3 second sample time.

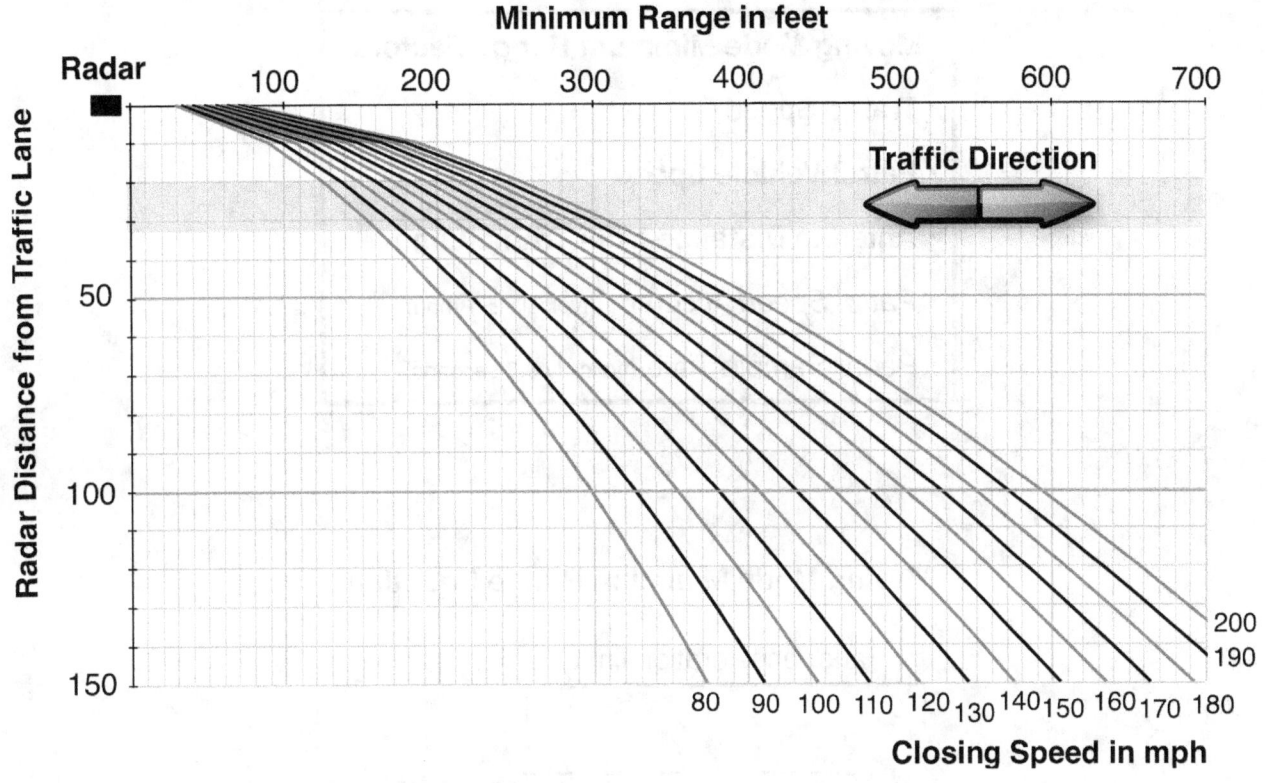

Figure 2.3-2 -- **MInimum Range**
**0.250 second Sample Time and ± 1 mph Accuracy**

---

**Moving Mode Minimum Range Factors**

· **Traffic Speed**

· **Patrol Vehicle Speed**

· **Antenna Distance from Traffic Lane**

· *Patrol Speed* Accuracy (typically ± 1 mph)

· Radar Sample Time (typically 0.3 seconds)

---

## Moving Mode Minimum Range Equation

radar acceleration limit

$$R_{min} = \sqrt{\frac{\left((v_o + v_p)\,d\right)^{4/3}}{\left(\dfrac{v_{acc}}{t_i}\right)^{2/3}} - d^2} + (v_o + v_p)\,t_i$$

Distance change during
1 sample period

$R_m$ = minimum range          $v_{acc}$ = patrol speed accuracy
$v_o$ = traffic speed          $t_i$ = radar sample time
$v_p$ = patrol vehicle speed   $d$ = antenna distance to traffic lane

# Chapter 3 -- Microwave Radar

## Chapter 3.1 -- Police Doppler Radar

Police microwave radars all use the Doppler Principle to measure speed. The radar transmits a continuous microwave signal and simultaneously measures the echo that is frequency shifted proportional to speed, the Doppler Shift. The frequency shift is measured in cycles per second and has unit dimensions of Hertz (Hz).

The horn from a moving train is a good example of the Doppler effect. As the train approaches a stationary listener the frequency pitch of the whistle sounds higher than when the train is even with the listener. As the train recedes from the listener the pitch decreases. Car horns exhibit the same phenomenon, as does all sound.

In the above example if a car horn is stationary and a listener is on the train the Doppler principle still applies. As the listener on the train approaches the stationary horn the pitch of the horn sounds higher, as the train recedes from the stationary horn the pitch sounds lower to anyone on the train.

## Stationary Radar

Microwave signals travel at the speed of light but still obey the Doppler Principle. Microwave radars receive a Doppler frequency shifted reflection from a moving object. Frequency is shifted higher for approaching objects, and lower for receding objects. The frequency shift is proportional to speed.

### Echo Frequency Shift

Approaching Traffic        Receding Traffic

$$f_t = f_o + f_d \qquad\qquad f_t = f_o - f_d$$

$f_o$ = radar transmit frequency        $f_t$ = traffic echo frequency
$f_d$ = Doppler frequency shift

## Frequency

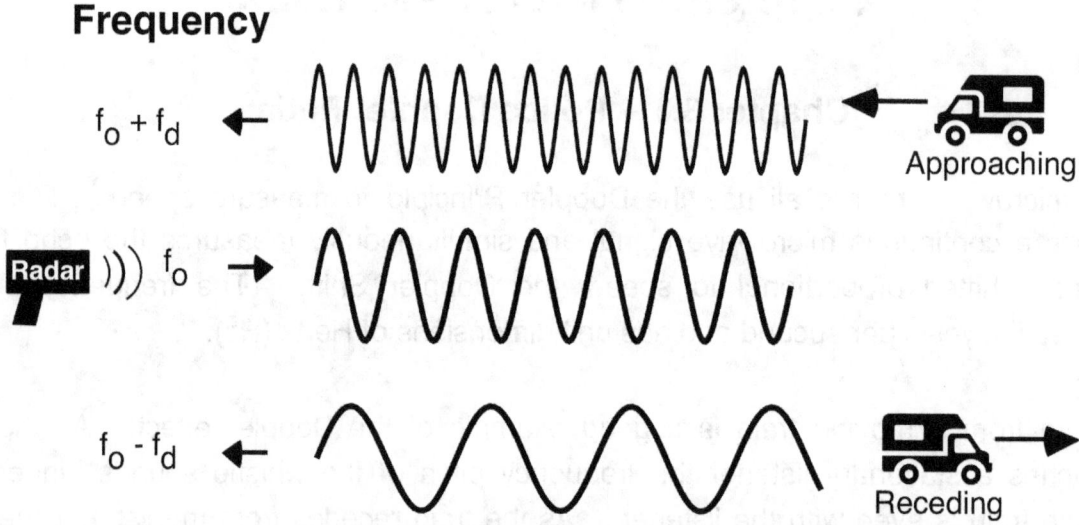

Figure 3.1-1 -- **Stationary Radar Doppler Shift**

Radar Doppler shift ($f_d$) is a function of transmit frequency ($f_o$), speed of wave (c = speed of light), and target vehicle speed (v). Speed is positive (+v) for approaching traffic and negative (-v) for receding traffic.

### Radar Doppler Shift Equation

$$f_d = \frac{2\,v\,f_o}{c}$$

$f_d$ = radar Doppler shift          v = vehicle speed
$f_o$ = transmit frequency          c = speed of light

A moving vehicle with a radar detector will receive a radar signal Doppler shifted proportional to speed.  The radar will receive a reflection shifted *again*, proportional to speed.  The echo the radar receives has twice the Doppler shift the target vehicle radar detector receives.

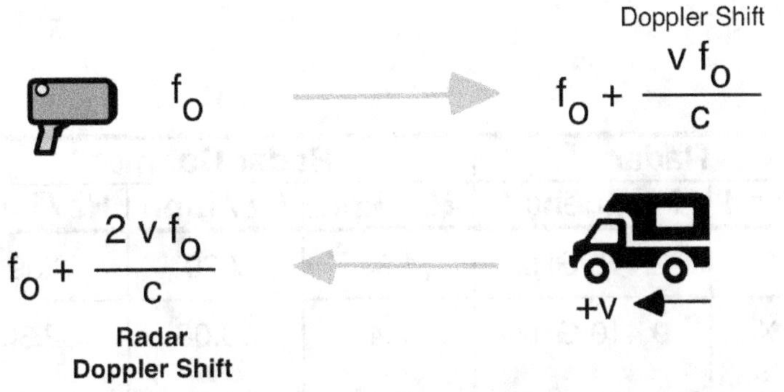

Doppler Shift

$$f_o + \frac{v\,f_o}{c}$$

$$f_o + \frac{2\,v\,f_o}{c}$$

**Radar
Doppler Shift**

$f_o$ = Transmit Frequency     v = Traffic Speed (-v for receding)
$f_d$ = Radar Doppler Shift     c = speed of light

Figure 3.1-2-- **Approaching Vehicle Doppler Shifts**

**Measured Speed**

$$v = \frac{c\,f_d}{2\,f_o}$$

v = Measured Speed - negative (-) for receding traffic
$f_d$ = Radar Doppler Shift - negative (-) for receding traffic
$f_o$ = Radar Transmit Frequency
c = Speed of Light

## Radar Doppler Shift at Police Frequencies

The table below shows radar Doppler shift per speed unit (1 unit) for common police radar frequencies.  Radar Doppler shift is **unit shift** multiplied by **speed**.  For example an X band radar Doppler shift for a 100 mph vehicle is 31.39 Hz/mph multiplied by 100 mph and equals 3,139 Hertz.

| Radar | | Radar Doppler | | |
|---|---|---|---|---|
| Band | Frequency | Hz / kph | Hz / mph | Hz / knot |
| S | 2.445 GHz | 4.53 | 7.29 | 8.39 |
| X | 9.410 GHz | 17.44 | 28.06 | 32.30 |
| X | 9.900 GHz | 18.34 | 29.53 | 33.98 |
| **X** | **10.525 GHz** | 19.50 | **31.39** | 36.12 |
| Ku | 13.450 GHz | 24.92 | 40.11 | 46.16 |
| **K** | **24.125 GHz** | 44.71 | **71.95** | 82.80 |
| **K** | **24.150 GHz** | 44.75 | **72.02** | 82.88 |
| **Ka** | **33.4 GHz** | 61.89 | **99.61** | 114.63 |
| | to **36.0 GHz** | 66.71 | **107.36** | 123.55 |

Table 3.1-1 -- **Unit Speed Doppler Shift**

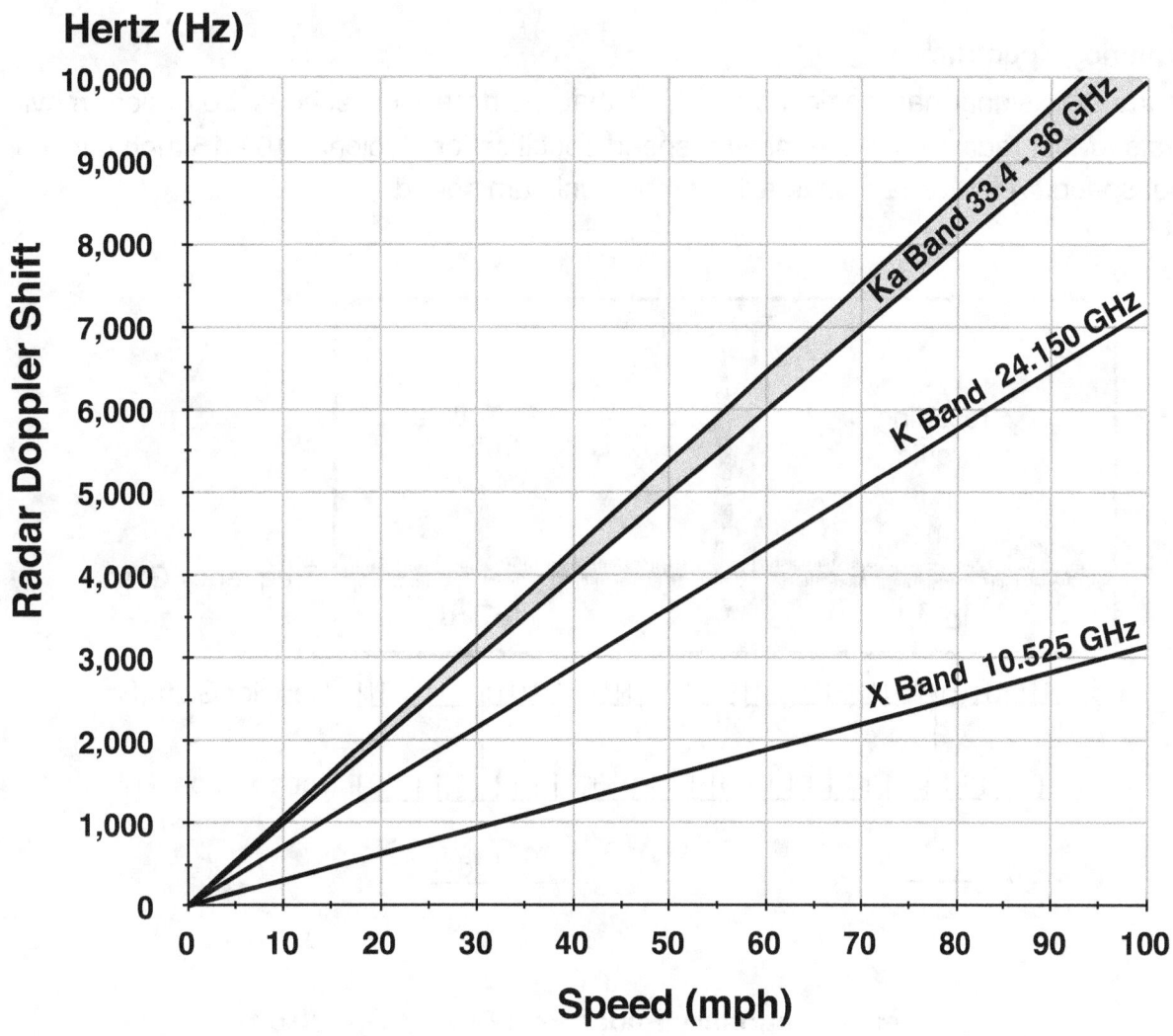

Figure 3.1-3 -- **Radar Doppler Shift vs Speed**

## Frequency Spectrum

The transmit signal has some noise skirts that interfere with echoes from slow moving objects. Most radars have a minimum speed specification, typically 10 - 15 mph minimum target speed. A few radars have a 1 - 3 mph minimum speed.

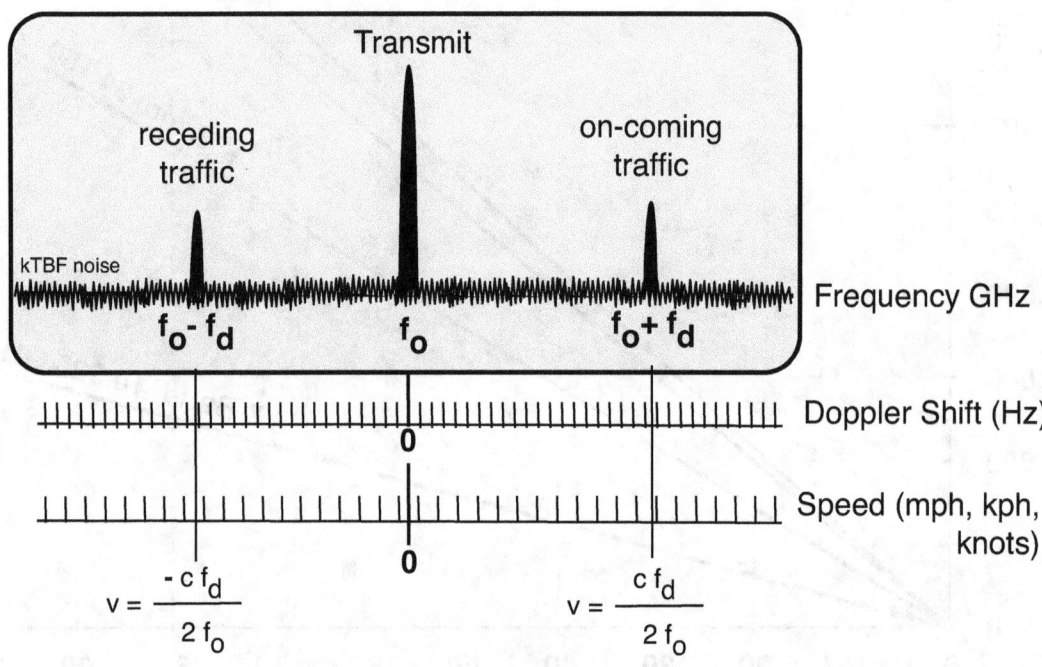

Figure 3.1-4 -- **Stationary Radar Frequency Spectrum**

Ground echo clutter is usually the strongest signal reflection, but because the ground is not moving it is does not have a Doppler shift. With moving-mode radar the ground echo is frequency shifted by the speed of the patrol vehicle, and so are all other reflections except those at the same speed as the radar.

kTBF is the average receiver noise level and is a function of **temperature** (T), radar receiver **noise bandwidth** (B), receiver **noise factor** (F), and Boltzmann's constant (k).

## Moving-mode Radar

Moving-mode radar is slightly more complicated. An on-coming target vehicle's echo is the sum of the patrol vehicle speed plus target vehicle speed - closing speed. The radar subtracts patrol vehicle speed from the echo to get target vehicle speed.

Moving-mode radar depends on two measurements to derive traffic speed:

1.) **Ground Echo** -- to determine patrol vehicle speed
2.) **Traffic Echo** -- Speed relative to moving radar

Moving mode opposite direction on-coming traffic echoes are the sum of patrol vehicle plus target vehicle speeds. Moving mode same direction traffic echoes are the difference of patrol speed minus target speed.

| Opposite Direction Traffic | Same Direction Traffic |
|:---:|:---:|
| $v_R = v_p + v_t$ | $v_R = v_p - v_t$ |

$v_t$ = traffic actual speed          $v_R$ = traffic speed relative to radar

$v_p$ = patrol speed

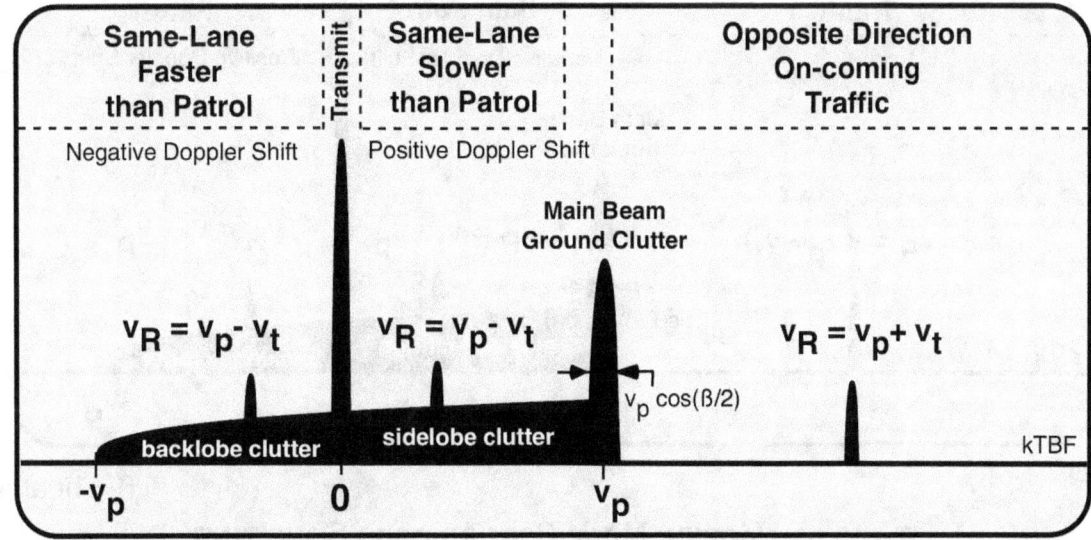

$ß$ = beamwidth

Figure 3.1-5 -- **Moving Mode Front Antenna Spectrum**

The transmit signal interferes with same-direction traffic traveling near the patrol speed. Same-lane mode cannot measure traffic traveling within ± 2 to 5 mph of the patrol vehicle.

The ground reflection interferes with slow moving same-direction and opposite direction traffic. Moving mode cannot measure traffic traveling slower than 15 to 20 mph.

### Rear Antenna

A radar with a rear antenna has negative Doppler's. Moving mode opposite direction receding traffic echoes are the negative of the sum of patrol plus target vehicle speeds. Same direction traffic echoes are the difference between target and patrol speeds.

**Moving Mode Rear Antenna**

Receding Traffic                    Same Direction Traffic

$V_R = -V_p - V_t$                    $V_R = V_t - V_p$

$V_t$ = traffic actual speed          $V_R$ = traffic speed relative to radar
$V_p$ = patrol speed

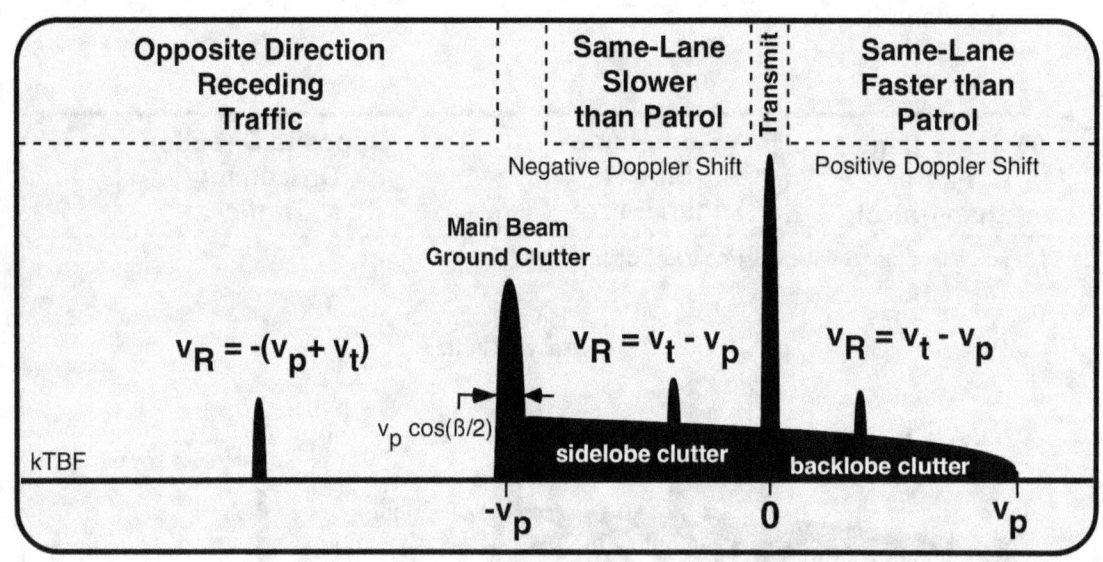

Figure 3.1-6 -- **Moving Mode Rear Antenna Spectrum**

Moving mode ground clutter consists of reflections from the ground and any stationary objects. Ground reflections come in from all angles, strongest reflections from the main beam, weaker reflections from the side and back antenna lobes. The reflections are Doppler shifted between plus and minus patrol speed ($\pm v_p$). Regions clear of ground clutter are limited by radar receiver noise and background interference and noise.

The largest and strongest ground clutter is in the main beam with a Doppler spread of patrol speed multiplied by the cosine of half beamwidth ($v_p \cos (ß/2)$ to $v_p$). The faster the patrol speed and the larger the antenna beamwidth, the wider the speed spread.

Moving mode reflections are Doppler shifted by traffic speed relative to patrol vehicle speed.

### Doppler Shift and Speed for On-coming Traffic

$$f_d = \frac{2 f_o (v_p + v)}{c} \qquad\qquad v = \frac{c f_d}{2 f_o} - v_p$$

$f_d$ = Radar Doppler Shift  
$f_o$ = Radar Transmit Frequency  

$v_p$ = Patrol Vehicle Speed  
$v$ = Target Vehicle Speed  
$c$ = Speed of Light

SPEED TERM SIGNS

The target and patrol vehicle speed terms ($v$ and $v_p$) are the negative of the terms in the above equations for some scenarios. *Front antenna same direction traffic* uses a negative speed term ($-v$). *Rear antenna receding traffic and patrol* speed use negative terms ($-v$ and $-v_p$).

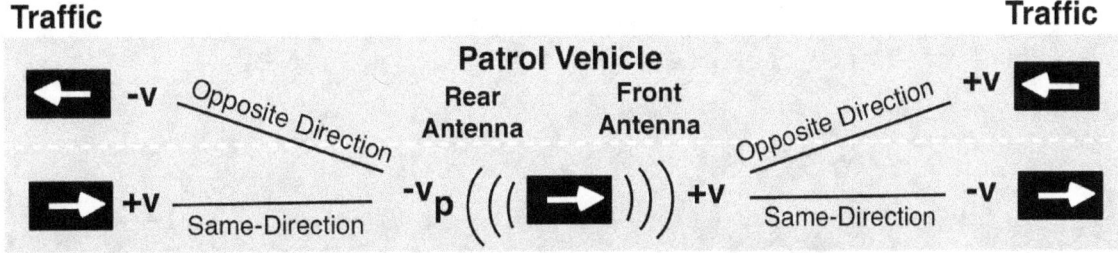

Figure 3.1-7 -- **Moving Radar Relative Speeds**

NOTES

# Chapter 3.2 -- Antenna Beam

## Antenna Beamwidth

Antenna beamwidth limits radar ability to distinguish multiple vehicles. When several vehicles are in the beam the operator must interpret, sometime guess, which vehicle the radar is tracking.

Beamwidth varies with model from about 9° to 25°, the larger the antenna and the higher the frequency the more narrow the beam. Circular antennas have a symmetrical beam, the same beamwidth in the vertical and horizontal planes. Rectangular and elliptical antennas have a more narrow beamwidth in the plane the antenna is wider, and a wider beamwidth where the antenna is more narrow.

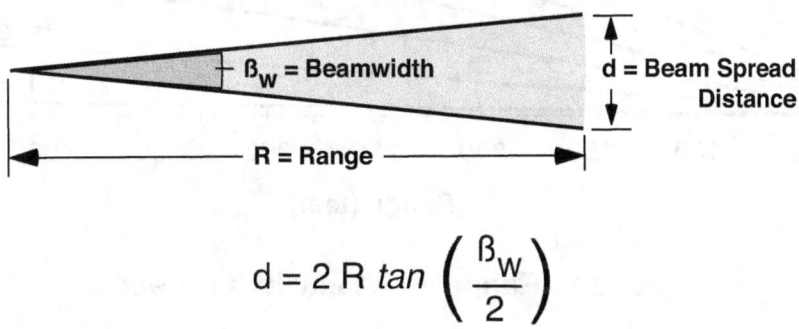

$$d = 2R \ tan \left( \frac{\beta_w}{2} \right)$$

Figure 3.2-1 -- **Beamwidth Spread**

d = beam spread distance
R = range
$\beta_w$ = beamwidth

The figure below graphs beamwidth spread in feet versus range for a beamwidth of 1, 5, 10, 15, 20, 25, and 30 degrees.  Most police down the road radars have beams between 9° and 15° that easily covers several lanes of traffic at relatively short ranges.

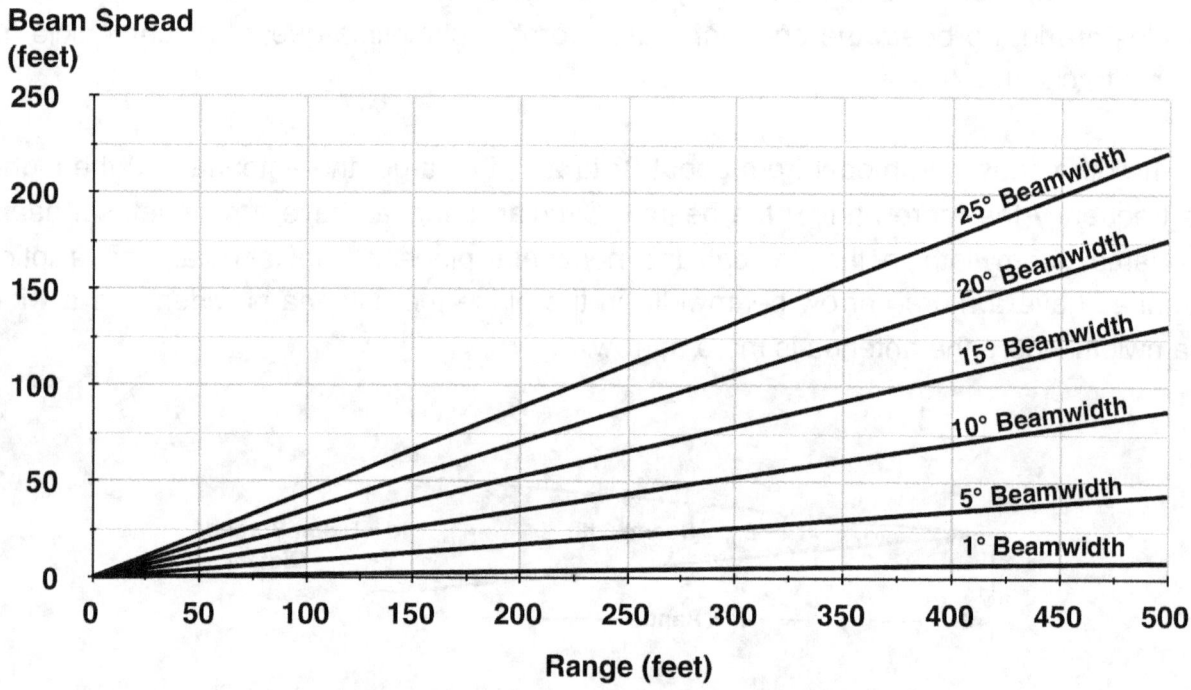

Figure 3.2-2 -- **Range vs Beamwidth Spread**

Antenna beamwidth is defined as the angle between the half power, -3 dB, points.  A radar has maximum detection performance in the beam center.  At the beam edge, -3 dB point, radar detection range decreases 29%. Detection capability drops dramatically past the beam edges, but not to zero.  Antennas have some detection capability outside the main beam in the antenna side lobes.

## Beam Detection Pattern

At the beam edges detection is down to 71% (-29%) compared to beam center. The vertical axis on the below chart is on a log scale for better resolution. The side lobes get a littler wider with lower magnitude as angle increases. The back lobes are unpredictable but still large enough for some detection capability.

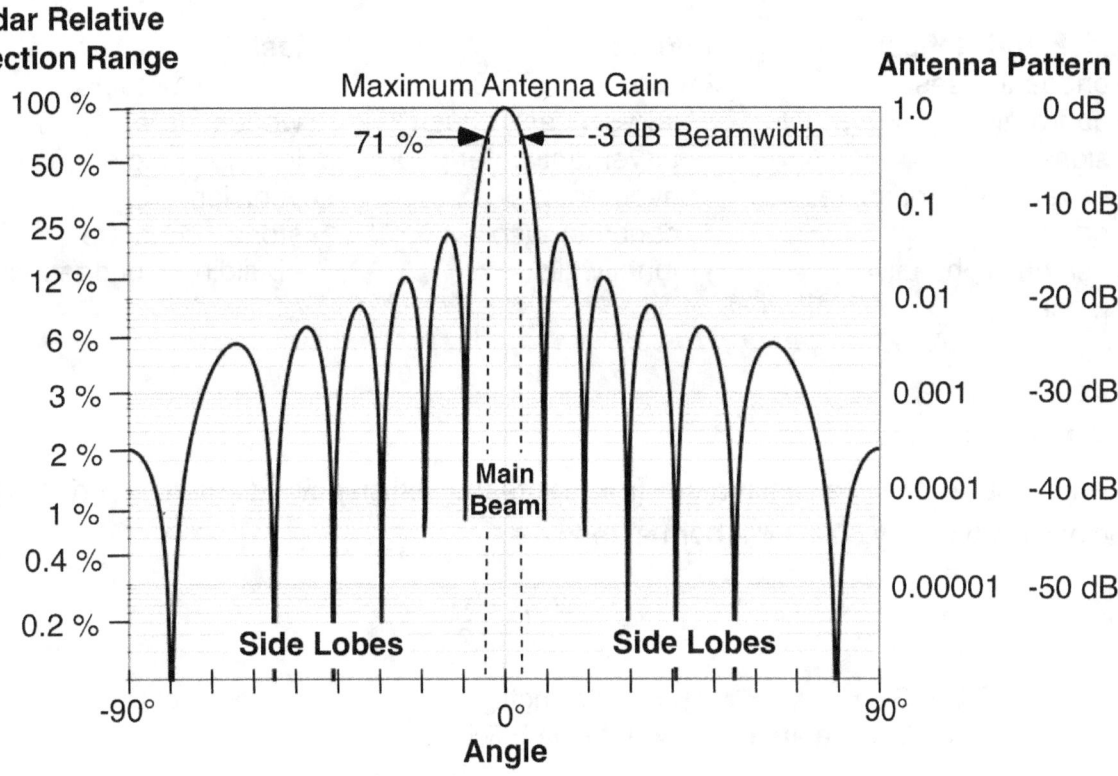

Figure 3.2-3 -- **Antenna Pattern vs Detection Range**

The antenna pattern is closely approximated by the *sin* x / x function.

$$f(x) = \frac{\sin x}{x} \qquad x = \pi \frac{n\,d}{L} \sin ß$$

d = antenna diameter        ß = angle off boresight (center)
L = wavelength              *n* = antenna efficiency
π = 3.14159

## Beam Masking

At microwave frequencies signal transmission and reflections are of a line-of-sight nature and do not bend or fringe around objects as do commercial AM and FM radio waves. Reflective objects mask and reflect radar signals, large non reflective obstructions can absorb and disperse signal energy.

### Objects that Mask or Reflect Microwave Energy

**Mask and Reflect**
- bridge trusses
- guardrails
- signs
  - overhead / roadside
- poles
  - signal, light, utility
- fences

**Reflect**
- water
  - lakes, rivers, ocean
  - pavement water pools
- pavement
  - concrete, asphalt
- tunnel walls

**Mask**
- hills / mountains
- woods / forest
- foliage / jungle
- vehicles
- stone / concrete
  - buildings, bridges, pillars

## Antenna Equations

Antenna **gain** can be estimated from antenna efficiency, diameter and frequency. Efficiency is typically 50% for a horn antenna.

$$G = 4\pi \frac{n A f^2}{c^2}$$

G = Gain          L = Wavelength          c = Speed of Light
A = Antenna Area     $n$ = Antenna Efficiency

A **circular horn** antenna **beamwidth** can be estimated from diameter and frequency.

$$\beta_w = 1.25 \frac{c}{f_o d}$$

$\beta_w$ = Beamwidth in radians          c = Speed of Light
d = Antenna Diameter                 $f_o$ = Frequency

### Beamwidth in Degrees = 845.32 / ( $f_o$ d )

$f_o$ = Frequency in GHz
d = Antenna Diameter in inches

# Chapter 3.3 -- Radar Configurations

## Configurations and Modes

Microwave radars come in two basic configurations, hand held and mounted, with various modes and options.

|  | **Hand Held** | **Fixed Mounted** |
|---|---|---|
| **Modes** | Stationary Mode, Moving Mode optional* | Stationary and Moving Modes |
| **Power Source** | Internal Batteries and/or Vehicle Battery DC plug. | Vehicle Battery |
| **Remote Control** | Option | Standard - Wired or Wireless |
| **Antenna Location** | | Antenna mounted inside vehicle. Outside Antenna optional. |
| **Second Antenna** | | Optional Front and Rear Antennas. |
| **Radar VSS cable option** | | Vehicle Speed Sensor (VSS) used for detecting patrol speed shadowing and speed bumping. |
| **Video cable option** | | Superimposes radar measured speed(s) on patrol vehicle video system. |
| **Frequency Band** | **X** Band, **K** Band or **Ka** Band | |
| **Transmissions** | *Continuous* and/or *Instant On* | |

* Hand held radars in moving mode require a fixture to securely mount the radar.

Table 3.3-1 -- **Radar Configurations**

## ANTENNA MOUNTING

Fixed mounted antennas should be located as high as possible and **must be aligned to patrol vehicle** direction of travel. Front mounted antennas should be directly facing the front of the vehicle, rear antennas facing aft. Any misalignment may result in high target vehicle speed readings in moving mode.

Front antennas are usually located on the dash board, right, middle or left. Rear antennas are mounted on the rear dashboard or one of the upper corners of the back windshield. Outside antennas are mounted in any of the upper corners by the passenger compartment roof.

## VSS CABLE

The VSS cable option allows the radar to compare radar measured patrol speed to the vehicle speedometer. If radar and vehicle measured speeds do not closely match, generally due to patrol speed shadowing or bumping, the radar stops processing target echoes until the speeds match. This prevents the radar from displaying what would be a high target vehicle speed reading.

## PATROL SPEED BLANKING

All moving radars are equipped with a *Patrol Speed Blanking* switch. Occasionally the ground echo patrol speed measurement gets stuck on off angle reflections forcing a low measured patrol speed. The patrol speed blanking switch breaks then reestablishes lock and usually corrects the problem.

## RADAR MODES

|  | **Stationary Mode** | **Moving Mode** |
|---|---|---|
| **Standard** | On-coming Traffic | Opposite Direction Traffic |
| **Optional** | Receding Traffic | Same Lane/DirectionTraffic |
| **Optional** | Fastest Target Mode<br>Radar displays strongest and fastest echo speeds. | |

Table 3.3-2 -- **Radar Modes**

# Displays, Status and Fault Indicators

## SPEED WINDOWS

Speed is sometimes displayed with a plus "+" or minus "-" sign. A plus sign indicates displayed speed is from an approaching vehicle, a minus sign indicates vehicle receding.

Some hand held radars use a single speed display window for measuring and locking speed readings. Some models have two speed windows, one for real time speeds and the other for locked speeds. The locked speed window is also used to display fastest vehicles for radars that can display strongest and fastest targets.

 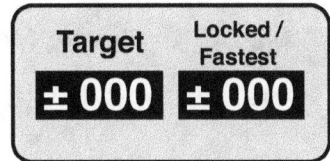

Figure 3.3-1 -- **Typical Hand held Stationary Radar Display**

Moving mode radars also display radar measured patrol vehicle speed.

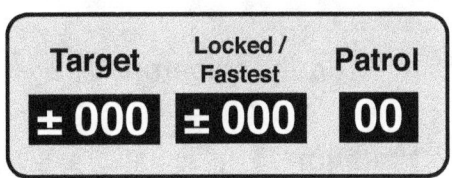

Figure 3.3-2 -- **Typical Moving Mode Radar Display**

## SPEAKER

Most radars have a speaker allowing the operator to hear an audio tone proportional to the Doppler frequency of the target vehicle. The faster the target vehicle the higher pitch the tone. Some models can sound an alert beep when measured speed exceeds a preset limit.

## Status Indicators

A radar may have separate status light indicators, or may display status in the speed window.

### Status Indicators

| Status | Mode |
|--------|------|
| **XMIT** | Transmitting |
| **STBY** | Standby |
| **Mov** | Moving Mode |
| **Sta** | Stationary Mode |
| **SL** | Same-lane moving Mode |
| **Fast** | Fastest Target Displayed |
| **Lock** | Target Speed Locked |

### Fault Indicators

| Fault | Condition |
|-------|-----------|
| **RFI** | Radio Frequency Interference present |
| **BIT Err** | Built-In-Test Error |
| **Xmit Err** | Transmitter Frequency or Power problem |
| **Low Volt** | Low Voltage condition |

A radar does not process targets if any fault error is indicated.

## Motorcycle Operation

Most motorcycle radars are hand held models intended for stationary operation only. The motorcycle should not be running and all radios, voice and data, should not be transmitting. When a violation is measured the radar is placed in STANDBY or OFF mode, put in a secure place such as a holster mount on the frame, and the operator starts the motor and proceeds after the speeder.

A few radar models physically mounts the radar to the motorcycle frame. The antenna beam is fixed making stationary operation more difficult to set up, but makes moving mode operation possible. There are a number of problems operating fixed mounted traffic radar from a moving motorcycle.

Police motorcycles have stiffer suspensions and a rougher ride than a four wheeled vehicle. Additionally motorcycles have much more engine vibration. The bumpier ride and engine vibrations can cause a speed bumping or batching effect and false readings. Also the antenna is mounted relatively close to the ground and favorable to patrol speed shadowing resulting in high target speeds.

Radar operation from a moving motorcycle is not CPL approved.

# Chapter 3.4 -- Test and Calibration

## Operator Testing

To insure radar hardware is setup and functioning properly the operator should perform several actions.

| Static Test and Checks | | |
|---|---|---|
| 1 | Check Radar Calibrated. | Most states require a radar be tested by a certified shop  periodically, typically once or twice a year.<br><br>A sticker on the radar, or records, should indicate last calibration test, next (due) test, and who tested. |
| 2 | Check Tuning Fork(s) Calibrated. | Same requirements as radar calibration. |
| 3 | Run Radar Self-Test. | Self test should be run before, during, and after use. |
| 4 | Check Radar with Calibrated Tuning Fork. | 2 tuning forks with different resonances required to test moving modes. |
| **Test Conducted at Operational Site** | | |
| 5 | Test Radar against Vehicle of Known Speed. | Test vehicle should have a calibrated speedometer for a valid test. Best to run test at operational location. |
| 6 | Test for Interference. | Set radar to *Receive Only* mode and scan for interference at operation site. |
| 7 | Set Range Control Setting. | Really receiver sensitivity setting. Low setting (long range) may allow unwanted interference. |

Table 3.4-1 -- **Operator Tests**

## Self-test

The degree a radar self-test and automatically adjust circuits varies with model and ranges from none or little to testing 50 percent or more of the electronics.  Self-test only checks and adjusts a portion of the electronics, the radar should also be checked with calibrated tuning forks before use.

**Radar self-test runs;**
- on power-up
- when the operator initiates
- automatically - periodically on the order of minutes
- automatically after a speed is locked
- any or all of the above

**Tests / Automatic Adjustments may include;**
- transmit frequency
- supply voltage (vehicle or internal battery)
- a simulated test target signal
- portions of the digital circuits
- portions of the transmitter / receiver
- display indicators
- any or all of the above

## Tuning Forks

Tuning forks are used to quick check radar speed accuracy.  A radar can measure a vibrating tuning fork and produce a speed reading proportional to the resonant frequency of the fork.  **Tuning fork resonance frequency equals radar Doppler shift**.

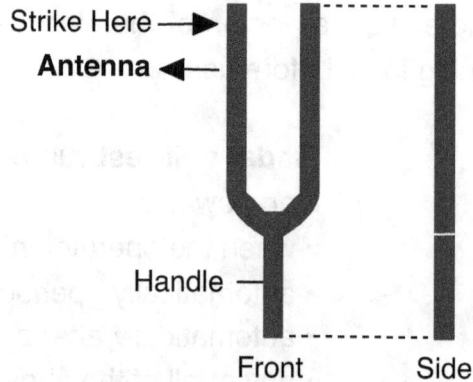

$$v = c \, f_d / 2 \, f_o$$

$v$ = speed
$f_d$ = Tuning Fork Resonance
$f_o$ = Radar Frequency

$$v = 0.3353 \, f_d / f_o$$

$$f_d = 2.9823 \, v \, f_o$$

$v$ = speed in mph
$f_d$ = Tuning Fork Resonance in Hz
$f_o$ = Radar Frequency in GHz

Figure 3.4-1 -- **Tuning Fork Test**

To check a microwave radar with a tuning fork the radar must be transmitting and the fork must be vibrating.  To start the fork vibrating gently strike the top side against a hard object such as wood or plastic, not metal.  Once the fork is vibrating place it a few inches in front of the antenna.  The fork's "side" must be facing the antenna to register a speed reading.

Moving mode radar requires a minimum of two tuning forks with different resonating frequencies, one to simulate patrol vehicle speed and the other to simulate traffic speed. The target speed should register as the **difference between the forks** speed.  In opposite direction moving mode the lower fork speed reads as patrol speed.

Tuning forks should be checked periodically to insure resonance has not changed due to a nick or bend from mishandling, or the fork mislabeled in previous tests.  Every tuning fork should have documentation indicating fork resonance and tolerance over temperature, radar frequency and induced speed.

### Tuning Fork Calibration Data

| Resonance | Temperature | Radar Frequency | Registered Speed |
|---|---|---|---|
| 2616 Hz | 80° F<br>Correction Factor<br>-0.02 mph / °F | Ka Band<br>34.700 GHz | 25.3 mph |

## Laboratory / Shop Calibration

Internal radar test circuits cannot check everything requiring manual testing periodically, typically once or twice per year at least. Periodic testing of any measuring instrument is standard practice.

### Periodic Calibration Test

- Vertical and Horizontal Beamwidth

- Effective Radiated Power (ERP)

- Power Supply Voltage Upper and Lower Limits

- Transmitter Frequency and Power to Antenna

- Receiver Sensitivity and Saturation (dynamic range)

- Speed Measuring Range

All test should be documented and include date, test conducted, and results.  A calibration sticker (BY, DATE, DUE) should be attached to the radar, or in records, indicating who calibrated the unit and when, and the date next calibration due.  Any tuning forks used to check the radar should also be tested and labeled at the time the radar is tested.

# NOTES

# Chapter 3.5 -- Operational Problems

## Dense Traffic Scenario

All radars display the speed of the target with the strongest echo. The strongest echo may be the closest target, the biggest target, or something in-between. A vehicle's echo varies with **size**, **shape**, and **range** from the radar. A compact car close to the radar may have a smaller echo than a truck at a greater range.

Dense traffic makes it difficult, if not impossible, for the radar operator to distinguish which vehicle the radar is tracking at any given time. In fast moving dense traffic a different vehicle is displayed almost every radar update period, fractions of a second to seconds.

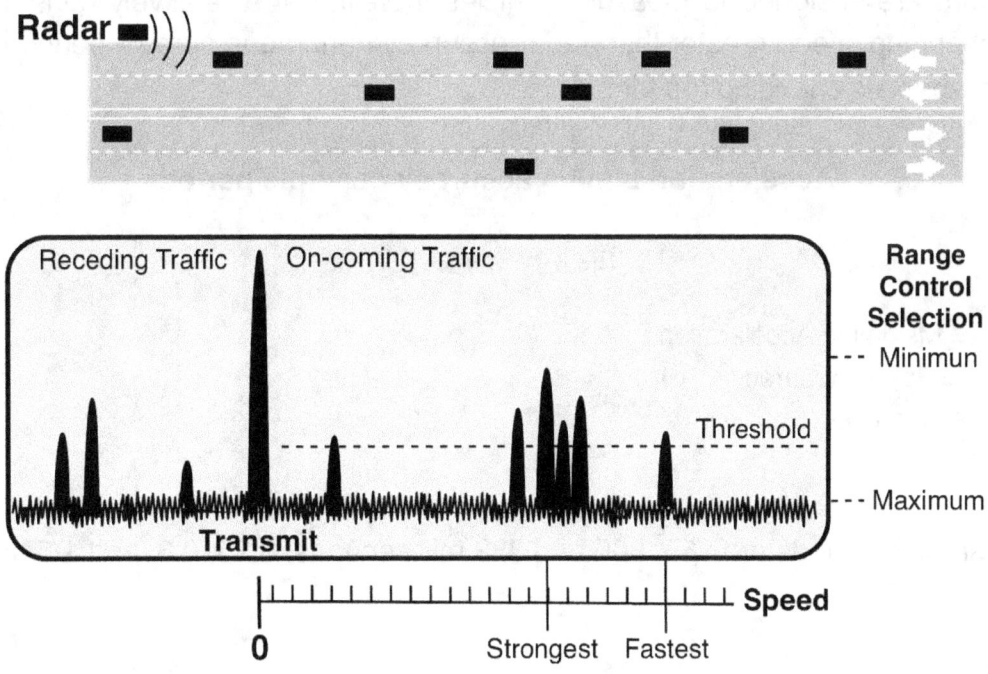

Figure 3.5-1 -- **Dense Traffic Scenario**

# Accuracy and Acceleration Limits
Applies to microwave and laser radars

## ACCURACY

Under ideal conditions most police radars are accurate to about ±1 mph. Microwave moving mode radar also measures *patrol vehicle* speed to an accuracy of about ±1 mph. Moving mode target vehicle accuracy is ±2 mph.

Some microwave and laser radars specify accuracy based on a percentage of vehicle speed. A typical specification is ±1 mph up to 60 mph, and ±1.6 % for speeds over 60 mph.

## ACCELERATION LIMITS

Police radars are designed to measure vehicles traveling at a relatively constant speed. Vehicles changing speed greater than radar accuracy during one sample period cannot be measured, speed is changing too fast.

### Acceleration Limit = accuracy / sample period

$$a_{max} = \pm\, v_{acc} / t_i$$

$a_{max}$ = Maximum Acceleration
$\pm v_{acc}$ = Speed Accuracy
$t_i$ = sample time

Common sample periods are 250, 300 and 333 milliseconds, 0.25, 0.3, and 1/3 seconds.

Table 3.5-1 -- **Acceleration Limit**

| Sample Period | | Change in Speed Limits |
| --- | --- | --- |
| seconds | milliseconds | |
| 0.167 | 167 ms | ± 6.0 mph / second |
| 0.2 | 200 ms | ± 5.0 mph / second |
| 1/4 | 250 ms | ± 4.0 mph / second |
| 0.3 | 300 ms | ± 3.3 mph / second |
| 1/3 | 333 ms | ± 3.0 mph / second |
| 1/2 | 500 ms | ± 2.0 mph / second |
| 1 | 1000 ms | ± 1.0 mph / second |

A radar or lidar cannot measure speed if the target vehicle is changing speed greater than the radar acceleration limits.  Both the cosine effect and traffic on curves introduce an acceleration component that will, if great enough, prevent the radar from obtaining a speed measurement.

## Instant-On Start-up Time

Instant-on radars are not exactly instantaneous, some turn-on or warm-up time is required. Police radars cut power to the transmitter diode until ready to measure. Once power is applied to the diode it takes typically less than a half a second, but can be as long as 2 seconds, to achieve steady state. Until steady state is reached any speed measurements are completely unreliable.

One manufacturer offers models with a pulsed mode, radar transmits a short 0.067 seconds burst that some radar detectors cannot pick up. The operator uses the pulsed mode to get an *estimate* of speed without setting off a radar detector. The operator cannot lock any speeds but must switch to normal transmissions, take another measurement and then lock a reading. The burst is so short that speed measurements are highly unreliable and useless.

Some agencies require minimum track times for a valid speed reading, typically 3 seconds for stationary operation and 5 seconds for moving mode.

## Strong Signals and Receiver Saturation

Strong echoes from large close vehicles will reduce radar detection range by automatically lowering receiver gain, automatic gain control (AGC), to protect the receiver from signal saturation. The AGC is a variable amplifier and is independent of the manual range control that sets receiver sensitivity threshold. Radar sensitivity and detection range is reduced unbeknownst to the operator.

Strong signals that saturate the radar receiver low noise amplifier or mixer will produce cross-modulation products, spurious signals and harmonics -- false signals. The interference will desensitize the receiver and can cause false speed readings. Moving mode radar has the potential to generate even more interference products because the ground echo, a very strong signal, is frequency shifted and another mixing source.

## Scanning Error

Scanning errors can occur when a hand held radar is pointed or scanned in a direction that picks up the patrol vehicle air-conditioner / heater fan or motor. Fan motion and motor electromagnetic noise can produce false speed readings. False readings are proportional to fan speed and reflection angle.

## Panning Error

Multi-unit radars will produce false readings when the antenna is panned around the display processor. Panning the antenna at different distances and angles produces different false speeds.

## Multi-Path

The transmitted signal does not always take a direct path, radar to target and back. , Overhead and off road highway signs can and do cause signals to take multiple paths in sometimes unpredictable ways.

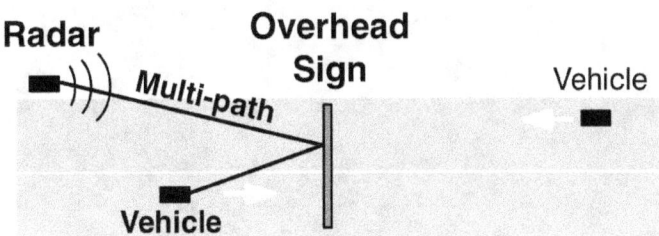

Figure 3.5-2 -- **Multi-path**

The above illustration is just one example of multi-path. A radar measuring on-coming traffic gets an overhead sign reflection from traffic receding the radar. The multi-path signal gets closer as the vehicle gets further in range, causing the Doppler shift to be positive like on-coming traffic.

NOTES

# Chapter 3.6 -- Vehicle Echo Size

## Radar Cross Section (RCS)

Radar cross section (RCS) is a measure of an objects' reflectivity, reflected energy divided by incident energy.  Most objects scatter most of the incident energy, reflecting only a small fraction back to the radar.  Large objects usually, but not always, have large reflections and a longer radar detection range.

The RCS dimension is in area, most commonly in square meters ($m^2$), or decibels with respect to 1 square meter (dBsm). Actual object area and RCS area use the same dimension, but are different concepts. RCS measures energy reflected based on incident energy and accounts for size, shape, directivity and reflectivity.

There are 6 major factors that affect an objects' RCS.

| Target | Radar Parameters |
| --- | --- |
| • Size | • Transmit Frequency |
| • Shape | • Signal Polarization |
| • Reflectivity | • Angle to Target |

### Shape

Flat surfaces reflect energy better than contoured shapes.   Edges and cavities with diameters on the order of wavelength tend to be good reflectors.   Shape can be as important as, or even more important than, physical size.

Some extreme examples include stealth aircraft such as the F-117 light bomber and B-2 bomber, reported to have RCS's on the order of the size of insects. The St. Louis Gateway Arch, a stainless steel 3 sided catenary curve structure 630 feet tall with a 630 foot base, is almost completely invisible to radar.

## Reflectivity

All metals and alloys are excellent reflectors at all frequencies. Metal screens with gap spacing less than a quarter wavelength are as reflective as a solid surface. At microwave frequencies carbon composites, fiberglass, and plastics are more transparent than reflective.

| Band - Frequency | 1/4 Wavelength |
|---|---|
| X - 10.525 GHz | 0.28 inches = 4.12 mm |
| K - 24.150 GHz | 0.12 inches = 3.10 mm |
| Ka - 33.400 GHz | 0.088 inches = 2.24 mm |
| Ka - 36.000 GHz | 0.082 inches = 2.08 mm |

Virtually everything is reflective, the degree of reflectivity varies with material and frequency.

## Frequency / Polarization

RCS depends on frequency and polarization. All electromagnetic waves have a polarization which is the orientation of the transmitted signal electric field, E-field. Common polarizations include vertical, horizontal, right hand circular, and left hand circular. Other polarizations include linear angled and elliptical. Police radars tend to use vertical or circular polarization.

Different frequencies or polarizations will produce different RCS patterns for the same object. Even ground reflections vary with frequency and polarization. In general the higher the frequency the larger the RCS.

## Angle

Most objects, spheres are an exception, reflect different amounts of energy for different angles. Many times when RCS is stated it will be an average of all important angles. Most automobiles have a larger RCS from the rear than the front.

## RCS examples

For a class of objects the larger the object the larger the RCS.  Objects that do the same tasks tend to be the same shape and made up of the same materials.  For physical size vehicles have relatively large RCS's compared to aircraft and watercraft.

| Vehicles | |
|---|---|
| **Vehicles** | |
| Pickup Truck | 200 |
| Car | 100 |
| **Aircraft** | |
| Jumbo Jet | 100 |
| Large Jet Airliner / Bomber | 40 |
| Medium Jet Airliner / Bomber | 20 |
| Large Fighter | 6 |
| 4 Passenger Jet / Small Fighter | 2 |
| Small Single Engine Aircraft | 1 |
| Conventional Winged Missile | 0.5 |
| **Watercraft** | |
| Cabin Cruiser | 10 |
| Small Pleasure Boat | 2 |
| Small Open Boat | 0.02 |
| **Wildlife** | |
| Large Bird | 0.01 |
| Insect | 0.00001 |

Table 3.6-1 -- **Radar Cross Sections in Square Meters ($m^2$)**

Source: Introduction to Radar Systems, Second Edition, Merril I. Skolnik, page 44, 1980.

At microwave frequencies average RCS can vary from 0.00001 $m^2$ for an insect to over 200 $m^2$ for a for a pickup truck.  Depending on frequency a man measures about 1 - 4 $m^2$. Weather, rain, sleet, hail, snow, fog, and clouds vary from about 100 - 10,000 $m^2$, and  are extremely frequency and polarization dependent.  Ground, clutter, varies from about 1,000 - 100,000 $m^2$.

## Automobile RCS

Typical automobile radar cross sections vary from about 10 to 100 or more square meters. Trucks are on the order of hundreds of square meters.

| Year Make Model | X Band - 10 GHz | | Ka Band - 35 GHz | |
|---|---|---|---|---|
| | Front | Rear | Front | Rear |
| 1970 Dodge Pickup | 200 | 300 | 84 | 670 |
| 1972 Oldsmobile Cutlass | 100 | 10 | 110 | 73 |
| 1972 Dodge Dart | 100 | 100 | 290 | 110 |
| 1970 AMC Gremlin | 200 | 32 | 84 | 73 |
| 1965 Ford Mustang | 10 | 100 | 55 | 42 |
| 1966 Chevrolet Corvette | 10 | 32 | 28 | 28 |
| 1967 Ford Cortina | 10 | 10 | 84 | 42 |
| Bicycle Coasting | 20 | | 2 | |
| Bicycle Pedaling | | 2 | | 7 |
| Man Walking | 1 | 1 | 3 | 4 |
| Corner Reflector | 100 | | 1300 | |

Table 3.6-2 -- **Measured Car RCS's in Square Meters (m²)**

Source: System Considerations for the Design of Radar Braking Sensors, IEEE Transactions on Vehicular Technology, vol VT-26, no. 2, page 151-160, May 1977.

The Corvette has a low RCS because a fiberglass body is much less reflective then a metal body. The Mustang convertible also has a low RCS in large part due to not having a metal roof. A vehicle's RCS is influenced by all the reflectors on the vehicle.

## Vehicle Reflectors

| Large Reflectors | Additional Reflectors |
|---|---|
| • Vehicle Body | • Head and Tail Lights |
| • Radiator / Grill | • Turn Signals |
| • Bumper | • Mirrors |
| • License Plate | • Driver and Passengers |
| • Antennae | |
|   - Radar Detector | |
|   - AM / FM | |
|   - Collision Avoidance Radar | |
|   - Vehicle Cell Phone | |
|   - Amateur Radio | |
|   - Business Radio | |

A radar detector increases a vehicle RCS because the detector antenna is a good radar reflector. Some police jammer antennas are reported to increase radar detection range 10% to 30%.

Some electrically heated windshields have metal film coatings reflective to microwave signals. These windshields will increase a vehicle's RCS dramatically, not to mention prevent the use of a radar or radar detector from behind the windshield.

## Estimated Vehicle RCS

| Vehicle Type | $m^2$ |
|---|---|
| Recreational Vehicle (RV) | 400 |
| Full Size Pickup Truck | 200 |
| Large Car | 120 |
| Mid Size Car | 60 |
| Small Car | 30 |
| Motorcycle | 10 |
| Bicycle | 5 |

For motorist with a vehicle that has a small RCS the good news is the vehicle is harder to track, the bad news is the vehicle may be mistaken for a more distant vehicle. Motorcycles are especially vulnerable because the RCS is so small compared to cars and trucks, on the order of 2 to 8 times smaller than a car, 13 times smaller than a pickup truck, and 27 times smaller than an RV.

## Multiple Vehicles

A vehicle with a large RCS could have the same echo signal power at a greater range than a close small vehicle. The range a distant large vehicle has the same signal return power or greater is a function of both vehicles' RCS and the range of the close small vehicle.

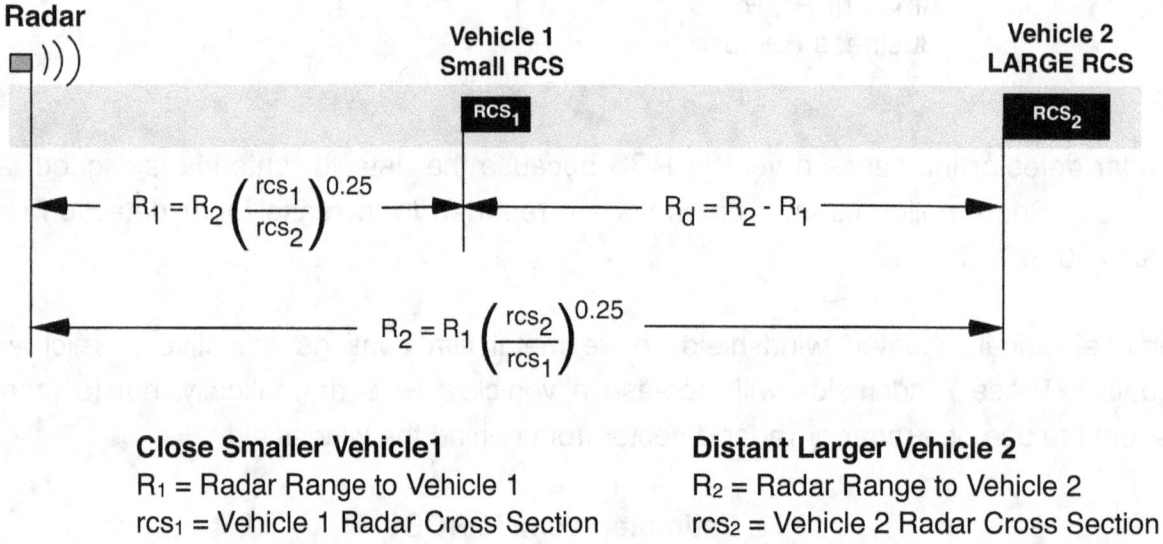

**Close Smaller Vehicle1**
$R_1$ = Radar Range to Vehicle 1
$rcs_1$ = Vehicle 1 Radar Cross Section

**Distant Larger Vehicle 2**
$R_2$ = Radar Range to Vehicle 2
$rcs_2$ = Vehicle 2 Radar Cross Section

Figure 3.6-1 -- **Vehicles Ranges for *Equal* Echo Power**

In the range equation everything cancels out except for range and the RCS's. The result shows the range difference is a function of range of close vehicle, and the vehicles' RCS ratio.

$$R_d = R_1 \left[ \left( \frac{rcs_2}{rcs_1} \right)^{0.25} - 1 \right]$$

The illustration below graphs the range difference ($R_d$) between 2 vehicles when the received signals from both vehicles are equal. The difference is a function of close small vehicle 1 range ($R_1$) from the radar and the ratio of Vehicle 2 RCS / Vehicle 1 RCS.

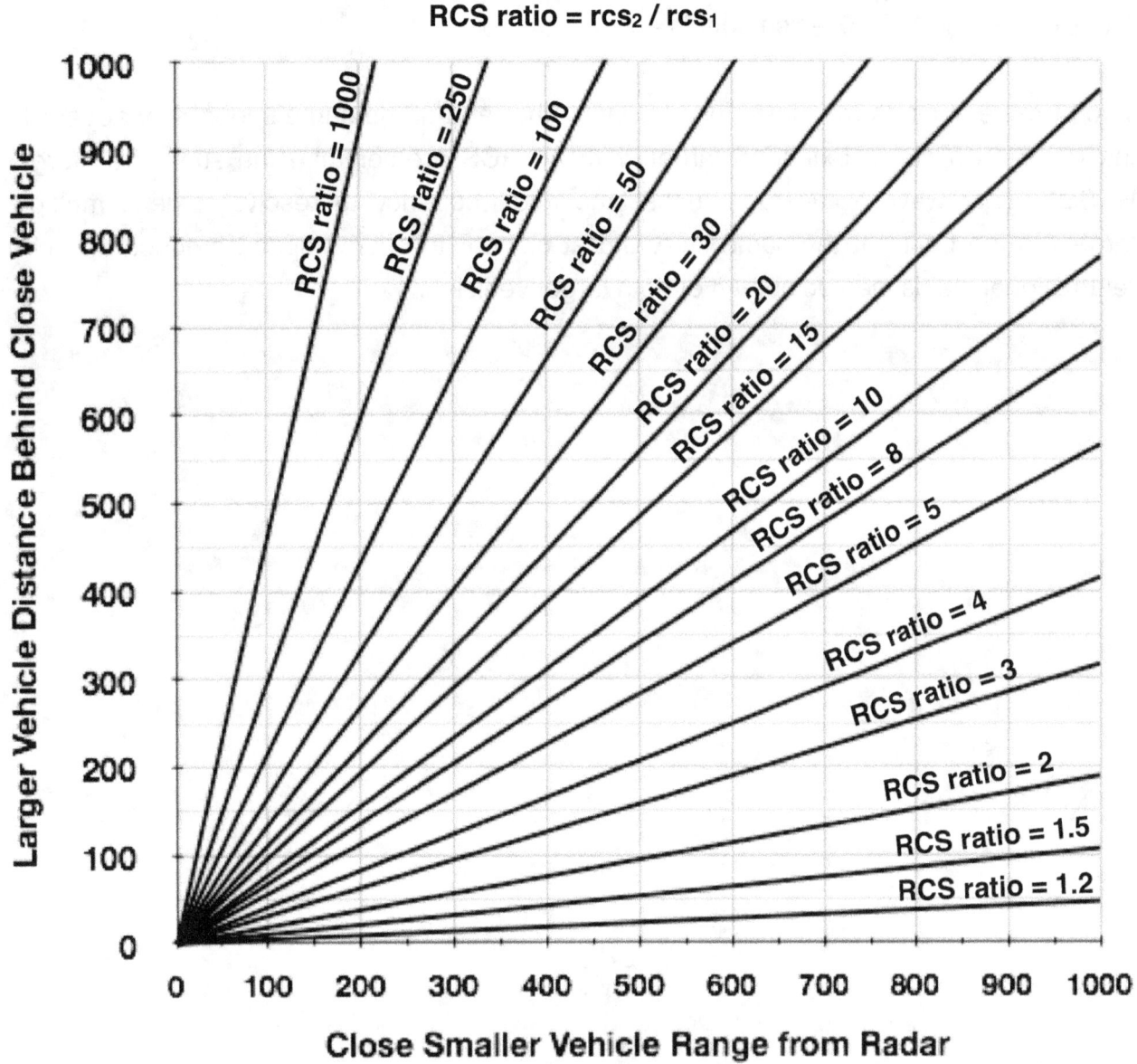

Figure 3.6-2 -- **Distance Between Vehicles when Echoes Equal**

Both range scales ($R_1$ and $R_d$) in the same dimensions (feet, yards, meters, etc.).

A radar 500 feet ($R_1$) from a small car with an RCS of 30 m² receives the same signal echo strength from a full size car with an RCS of 120 m² that is 210 feet behind the small car. The RCS ratio is 120/30 which equals 4.

Two or more vehicles traveling close to each other and at the same speed may appear to the radar as a single extended target, with an RCS greater than the sum of all target RCS's. Microwave traffic radar uses time and frequency to resolve targets, multiple targets in the beam at the same time and at or near the same speed produce an echo return signal that is the vector sum of both target vehicles.

# Chapter 3.7 -- Moving Mode Problems

Moving mode radar measures the closing or opening traffic speed relative to the radar. Traffic speed is derived from patrol vehicle speed, any error in patrol speed translates directly to a traffic speed error. The radar measures patrol speed by measuring the ground speed Doppler shift **directly in front** of the patrol vehicle, or directly behind for aft antennas. Off angle ground reflections measure patrol vehicle speed low, *patrol speed shadowing* by definition. The greater off angle from patrol vehicle center line, the greater the patrol and target speed error.

## Patrol Speed Shadowing

Large reflective objects can cause the radar to measure patrol speed off angle. A misaligned antenna will also cause off angle ground reflections. Patrol speed measures low by the cosine of the angle, the cosine effect.

$$V_{pm} = v \; cos \; \beta$$

$V_{pm}$ = Measured Patrol Speed
$v$ = Actual Patrol Speed
$\beta$ = Ground Reflection Angle Error

In opposite direction moving mode a low patrol speed reading translates directly to a high target speed reading. The target speed reading is high by the speed the patrol measurement is low.

The error in same-lane mode depends whether the target vehicle is traveling faster or slower than the patrol vehicle. Target vehicles traveling faster than the patrol vehicle have a high speed error for a low patrol speed. Target vehicles traveling slower than patrol vehicle have a low speed error for a low patrol speed.

| Opposite Direction Traffic | Same Direction Traffic | |
|---|---|---|
| | Slower than Patrol | Faster than Patrol |
| Traffic Speed **High** by Patrol Speed Error | Traffic Speed **Low** by Patrol Speed Error | Traffic Speed **High** by Patrol Speed Error |

Table 3.7-1 -- **Traffic Speed Error when Patrol Speed Measures Low**

Patrol speed shadowing is not uncommon and can be caused by numerous objects and situations. Police radars have a *Patrol Speed Blanking* switch to break and reestablish ground lock if the radar gets stuck on an off angle ground echo.  Some radars connect to the patrol vehicle speedometer, if speedometer and radar measured speeds do not match speed readings are disabled.

| Objects that can cause **Extended Speed Shadowing** | Objects that can cause **Momentary Speed Shadowing** |
|---|---|
| • Sound / Construction concrete barriers | • Pillars |
| • Metal Guardrails | • Overhead Signs |
| • Crash Barriers - metal cables / post | • Road Signs |
| • Bridge Trusses | • Parked Vehicles |
| • Hill-cuts / Depressions / Tunnels | • Ice Patches, Snow Piles |
| • Highway Barrels - metal / water filled plastic | |
| • Ditch Water | |
| • Moving Vehicles | |
| • Multiple Parked Vehicles | |
| • Plowed Snow Piles | |

## Shadowing from Moving Vehicles

Traffic traveling in the same direction and slower then the radar can be mistakenly used by the radar as the ground reflection. Radar measured patrol speed will be the difference between the patrol and vehicle speeds.

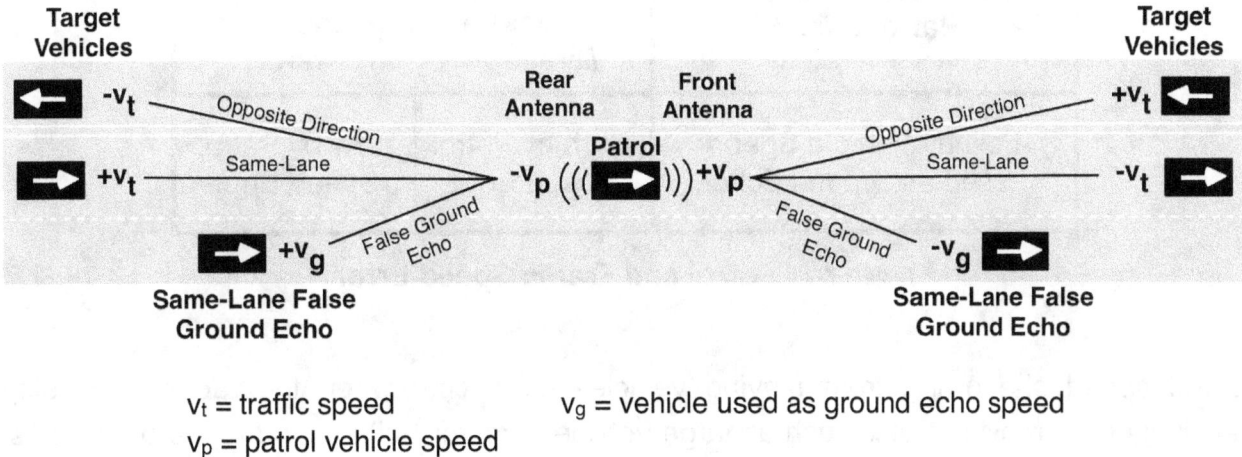

$v_t$ = traffic speed            $v_g$ = vehicle used as ground echo speed
$v_p$ = patrol vehicle speed

Figure 3.7-1 -- **Moving Mode Relative Doppler Speeds**

The below equations use relative Doppler speeds, speed is positive for a positive Doppler shift and negative for a negative Doppler shift.

| Traffic | Front Antenna | Rear Antenna |
|---|---|---|
| Opposite Direction | $V_{pm} = V_p - V_g$ | $V_{pm} = -V_p + V_g$ |
| | $V_m = V_t + V_p - (V_p - V_g)$ | $V_m = -V_t - V_p - (-V_p + V_g)$ |
| | $\mathbf{V_m = V_t + V_g}$ | $\mathbf{V_m = -V_t - V_g}$ |
| Same-Lane | $V_{pm} = V_p - V_g$ | $V_{pm} = -V_p + V_g$ |
| | $V_m = V_p - V_g - (V_p - V_t)$ | $V_m = -V_p + V_g - (V_t - V_p)$ |
| | $\mathbf{V_m = V_t - V_g}$ | $\mathbf{V_m = -V_t + V_g}$ |

$V_{pm}$ = measured patrol speed            $v_p$ = patrol vehicle speed
$V_m$ = measured traffic speed            $v_t$ = traffic speed
$v_g$ = vehicle speed used as ground echo

The below table summarizes the above equations for patrol speed shadowing from moving vehicles and applies to front and rear radar antennas.

| Measured Patrol Speed | Measured Traffic Speed | |
|---|---|---|
| | Opposite Direction | Same-Lane Traffic |
| **Low** by **Vehicle Speed** used as Ground Echo | **High** by Patrol Speed Error | **Low** by Patrol Speed Error |

Table 3.7-2 -- **Patrol and Traffic Speed Error**

Patrol speed shadowing from moving vehicles can occur when the radar is passing relatively slow moving traffic such as large vehicles, tractor-trailers, and a line of vehicles. Same direction vehicles traveling faster than the patrol vehicle will not be mistaken for a ground reflection because the Doppler shift is negative relative to patrol speed.

**Patrol Speed Shadow Management**

To manage shadowing from stationary or moving objects some radars monitor patrol vehicle speedometer (VSS - Velocity Speed Sensor). If the radar measured patrol speed does not closely match the speedometer target measurements are not displayed until speeds match. Speedometers are not accurate enough for radar to use in speed calculations, but a good check for patrol speed shadowing.

# Batching or Speed Bumping

Batching or Speed Bumping occurs when the patrol speed suddenly changes speed causing a false or inaccurate reading. Radar measured traffic speed and patrol vehicle speed are not updated simultaneously, if patrol speed changes suddenly the radar may still be using outdated patrol speed data leading to a momentary measured speed error. Sudden acceleration, hit the gas, braking, turning, curves, or hitting bumps can cause batching or speed bumping.

# Chapter 3.8 -- Interference

## Interference Types

Most of the time police radars can identify interference and adapt by adjusting gain or stop processing until the interference stops or lowers to a manageable level. The radar alerts the operator with a radio frequency interference (RFI) indicator and stops displaying speeds.

### Interference Effects

- **Reduces Radar Detection Range**
  Automatic gain control adjust out interference and detection range.

- **Activates Radar Interference Detection Circuit**
  Stops processing speeds when interference detected.

- **Mask Legitimate Targets**
  No indication to operator.

- **Produces False Speed Readings**
  No indication to operator.

Interference can cause the radar to stop processing or produce a false speed reading, sometimes out of nowhere. The radar's self adjusting gain to manage interference can significantly lower detection range unbeknownst to the operator, especially if radar range manual adjustment is set to maximum.

### Mutual Interference

**Two or more traffic radars transmitting** at on near the same frequency operating in close proximity **will interfere** with each other. The degree of interference depends on radiated power, antenna pointing angles, and distance between radars.

## Radio Frequency Interference (RFI)

Interference occurs when unwanted radio frequency energy gets into the radar circuits. Most interference gets into the radar receiver by the antenna. Strong energy sources can penetrate almost any circuit from any angle. High power transmitters generate noise well outside their transmitting band and adds to other background noise.

### Radio Frequency Transmitters

**Common Sources**

- Airports
  - -- radars (weather, tracking)
  - -- communications
  - -- beacons
- Communication Towers
  - -- TV /AM / FM broadcast
  - -- microwave / radio relays
  - -- police / fire services
  - -- cellular services
- Military Installations
- Hospitals
- Telephone / Cable antenna farms
- Amateur Radio
- Satellite Uplinks
- Weather Radars
- Aircraft / Marine Vessels
- Power Plants / Transmission Lines
  - -- generator sub-stations
  - -- transformer sub-stations
  - -- transmission lines
  - -- pole transformers

**Patrol Vehicle Sources**

- **100 watt** radio (VHF/UHF)
- **2 watt** walkie-talkies (VHF/UHF)
- High Frequency Radio (HF)
- **CB** Radio (HF)
- Cellular Phones (UHF)
- Data Communications System
- Satellite Uplink

## Field Disturbance Sensors

Field disturbance sensors are radio or microwave transceivers used for short range detection of objects or motion.  These sensors operate in multiple bands, including police radar X and K bands.  Sensors in the X and K bands are numerous and widespread and can interfere with police radar and radar detectors.

Table 3.8-1 -- **Field Disturbance Sensor Applications**

| Application | Use |
|---|---|
| **Intrusion Sensors Burglar Alarms:** | • commercial stores <br> • warehouses <br> • public buildings |
| **Automatica Door Openers:** | • grocery and retail stores <br> • truck and automobile entrances <br> • railroad train entrances |
| **Obstruction Detectors on vehicles:** | • fork lifts <br> • moving farm equipment <br> • railroad locomotives, rolling stock, specialized cars |
| **Object Detectors:** | • production line counting and detection |
| **Speed Measuring:** | • train yards <br> • sporting events <br> • police |

The Federal Communications Commission regulates field disturbance sensor frequencies and power in FCC Rules and Regulations, Part 15.

| Band | Frequency | Field Strength | Effective Radiated Power |
|------|-----------|----------------|--------------------------|
| X | 10.525 GHz ± 25 MHz<br>10.500 - 10.550 GHz | 2.5 volts / meter<br>at 3 meters | 2 watts ERP |
| K | 24.125 GHz ± 50 MHz<br>24.075 - 24.175 GHz | | |

Table 3.8-2 -- **Field Disturbance Sensor Limits**
Same as police radar X and K band limits.

Low power field disturbance sensors include unlicensed and unattended radars, and some sports radars. *Amateur Radio* primary frequencies are between 24.0 - 24.05 GHz, with secondary frequencies between 24.05 -- 24.25 GHz.

| Band | Frequency | Field Strength | Effective Radiated Power |
|------|-----------|----------------|--------------------------|
| K | 24.125 ± 100 MHz<br>24.025 - 24.225 GHz | 0.25 volts / meter<br>at 3 meters | 0.02 watts ERP |

Table 3.8-3 -- **Low Power Field Disturbance Sensor Limits**

Police K band radars also shares the frequency with;

- Amateur Radio,
- Fixed Point-to-Point Communications,
- Industrial, Scientific, and Medical Equipment.

## Natural and Unintentional Radiation

### Natural Interference

Temperature affects the sensitivity of all receivers, the lower the temperature the better the sensitivity. The sun affects background noise, as the sun heats objects the thermal excitation of atoms in conducting materials generates noise.

At X band atmospheric conditions have much less effect on signal propagation than at K or Ka band. Both K and Ka bands signals are attenuated by oxygen and water vapor in the atmosphere. Signals in K band are attenuated more by water vapor due the the spacing of the atoms that resonate at that frequency. Rain significantly reduces radar detection for K and Ka bands, much less reduction at X band.

Pointing a radar in the direction of rain drops will produce an average speed of the rain in the beam. Rain falls at speeds from about 14 mph for a light moderate rain to about 25 mph or more for a heavy rain.

### Unintentional Forms of Radiation

Unintentional sources of interference include automotive ignitions, alternators, spark plugs, and wiper motors, those on the patrol vehicle and on other vehicles close to the radar. The ignition draws a lot of current for a short time, the more current the higher the electromagnetic field levels. Alternators are electromagnetic devices by nature, fields vary with engine rpm (revolutions per minute). Spark plugs generate a great deal of broadband noise, fast and short high power pulses. Wiper motors generate strong electromagnetic fields that change with time at the wiper rate.

Fluorescent, mercury, sodium vapor (common for street lighting) and neon / argon lights switching on or off can trigger false radar readings. Switching high voltages produces high interference fields.

Power plants, generator sub-stations, transformer sub-stations, transmission lines and pole transformers are strong sources of interference. This type of interference usually causes a buzzing or humming of the radar audio Doppler and/or false speed readings.

# Chapter 3.9 -- Other Situations

## Marine Radar

Some agencies use police radar to measure watercraft speeds.  Marine radar must meet the same Federal Communication Commission standards as police radar.  Additionally marine radars operate in a harsh environment and must be hermetically sealed.

Marine radar is measuring much smaller radar size targets then land radar.  Trucks and cars are 10 to 10,000 times larger radar targets than boats of comparable size, largely due to shape and materials differences.  Land vehicles are typically 10 m² to 200 m² radar size, boats vary from about 0.02 m² to 10 m².  Radars measuring boats will not have the same performance as measuring land vehicles.

### Multi-path

Water is highly reflective, about 100 times greater than dry ground, aggravating multi-path problems.  Multi-path occurs when reflections return by indirect paths instead of directly from radar to target and back.  Multi-path can go from radar to water surface to target and back, or from radar to target back to radar via the water surface.  Multi-path is unavoidable on water.

Moving water reflections introduce an error proportional to water speed.  The measured speed is high when the water and target are in the same direction, and low when the water and target speeds are in opposite directions.  Water that is perfectly still and calm will not cause any speed errors but will cause the target reflection to fade in and out.

## Water Wave Speed

Water wave speed is a function of water wavelength, the longer the wavelength the greater the speed.  Water wavelength is the distance between wave peaks, or wave valleys.

WAVE TYPES
- Wind Waves
- Swell Waves -- generated outside area
- Gravity Waves -- typically greater than 2 inch wavelength
- Capillary Waves -- surface tension ripples typically less than 1 inch wavelength
-  other waves riding waves

### Classic Ocean-wave Theory for Speed

$$v_w = ( g\, L_w / 2\, \pi )^{0.5}$$

$v_w$ = Water Wave Speed      $\pi$ = 3.141592
$L_w$ = Water Wavelength      $g$ = Acceleration Due to Gravity

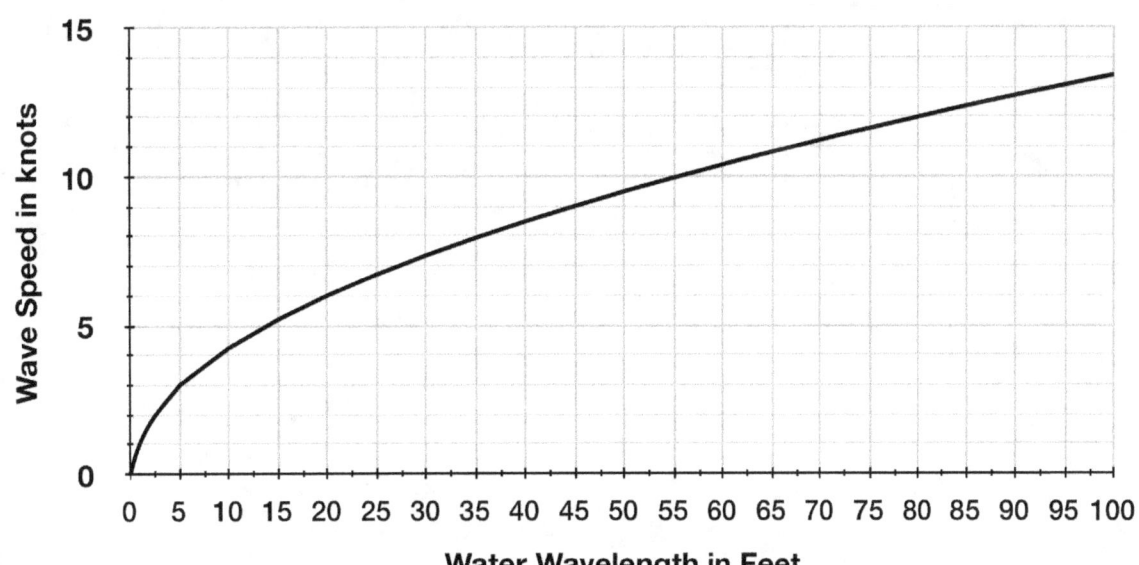

Figure 3.9-1 -- **Wavelength versus Wave Speed**

## Radar in a Tunnel

Microwave police traffic radar should not be operated inside a tunnel. These radars are designed to operate in free space, no obstructions and as high above ground as practical. Radar beam pattern in a tunnel is greatly distorted by the tunnel side walls and ceiling. Operating inside a tunnel virtually guarantees multi-path problems.

Strong reflections or interfering signals have the potential to saturate a radar receiver low noise amplifier and mixer producing harmonics and spurious signals and false targets. These false targets literally come out of nowhere, and can produce various false speed readings depending on speeds and echo amplitudes of all moving vehicles.

Radar operation in a tunnel is not CPL approved.

# Chapter 3.10 -- Department of Transportation Radar Tests

U.S. Department of Transportation (DOT)

National Highway Traffic Safety Administration (NHTSA)

*Police Traffic Radar ISSUE PAPER*

February 1980, DOT HS-805 254

## Test Summary

Seven radar units, six different models, from four different manufacturers were tested for problems. The study listed several recommendations based on problems encountered with some of the radars in the test. All the radar units tested operate at X band, two test models were handheld, five test models were multi-piece radars.

| Manufacturer | Model* | Test Radar |
|---|---|---|
| CMI - handheld | Speedgun 6 | D |
| CMI - handheld | Speedgun 8 | E |
| Decatur Electronics | MV-715 | C |
| Kustom Signals | MR-7 | G |
| Kustom Signals | MR-9 | A |
| M.P.H. Industries | K-55 | B and F |

Table 3.10-1 -- **Test Radars**

* From unofficial sources,.

The test data does not identify the exact radar model used but instead categorizes the radar units as A, B, C, D, E, F, and G. The test revealed several operational and multiple interference problems. Below is a summary of the test results.

| Radar | | | | | | | Percent | Problem |
|---|---|---|---|---|---|---|---|---|
| A | B | C | - | - | F | G | 100% | Panning Error (all multi piece units) |
| A | B | C | D | E | F |   | 86% | Patrol Speed Shadowing (moving mode) |
| A | B |   | D | E | F | G | 86% | Patrol Vehicle 100 Watt Radio |
| A | B | C | D |   | F | G | 86% | Patrol Vehicle CB Radio Transmissions |
|   | B | C | D | E | F |   | 71% | Scanning Error |
| A | B |   |   | E | F | G | 71% | 2 Watt Hand Held Radio |
|   | B | C | D | E |   |   | 57% | Patrol Vehicle AC/Heater Fan and Motor |
|   | B |   | D |   | F |   | 43% | Speed Bumping, Batching (moving mode) |
| A | B |   |   |   | F |   | 43% | External Police Radios 20 - 30 feet away |
|   |   | C | D |   |   |   | 29% | Ignition and Alternator |
|   |   |   |   |   | F |   | 14% | External CB Transmissions |
|   |   |   |   |   | F |   | 14% | Transmission Out of Band (+900 kHz) |

6   9   6   7   5   10   4   Total

Table 3.10-2 -- **Radar Problems**

Radars D and E (CMI Speedgun 7 and 8) were hand held, panning error does not apply.

Test radar F had the most problems (10), test radar B had the second most problems (9), both are believed to be the same model, M.P.H. Industries K-55.

**Tuning Forks**
Forty two percent, 5 out of 12, tuning forks tested were mislabeled or mis-calibrated.  Four were low by 1 mph, one was high by 1 mph.

Radar B traveling 40 mph   - tracked 37 mph Pinto
                           - lost track several seconds
                           - tracked car well behind Pinto

Radar D traveling 41 mph   - tracked 57 mph truck well behind 35 mph Pinto
                           - eventually picked up Pinto

Radar G traveling 41 mph   - tracked 57 mph truck 100 yards behind 33 mph Pinto

Table 3.10-3 -- **Moving Mode Abnormalities**

| | |
|---|---|
| Beamwidth: | 13.3° - 24.8° |
| Effective Radiated Power (ERP): | 26.3 - 134 milliwatts |
| Low Voltage Condition: | 6 - 11.9 Vdc |
| Maximum Detection Range: | 105 feet - 1.1 miles |

Table 3.10-4 -- **Performance Variations**

MAXIMUM DETECTION RANGE TEST

86 moving mode test conducted.
3 different target vehicles;

- 2 door Ford Pinto,
- 2 door Ford Thunderbird,
- Augmented Winnabego Mobile Home.

Two stationary mode test with one radar and one target vehicle, the Pinto.

| Radar: | A | B | C | D | E | F | G | Medium | Average |
|---|---|---|---|---|---|---|---|---|---|
| **Beamwidth:** | 13.3° | 20.4° | 17.5° | 18.8° | 18.6° | 24.6° | 14.3° | 18.6° | 18.2° |
| **ERP (mW):** | 85 | 134 | 120 | 34.3 | 39.2 | 26.3 | 56 | 56 mW | 71 mW |
| **Low Voltage:** | 11.9 | 6.0 | 7.6 | 10.2 | 10.3 | 6.6 | 11.9 | 10.2 Vdc | 9.2 Vdc |

mw - milliwatts

Table 3.10-5 -- **Beamwidth, ERP, and Low Voltage Summary**

## Recommendations

Based on the test results the report suggested several recommendations as summarized below.

- Adopt **radar standards** and require police agencies to follow standards.
- Develop policy guidelines for radar **maintenance, testing, and calibration**.
- Keep adequate **maintenance and calibration records**.
- Establish minimum **training standards**.
- Develop **State-level certification** that must be renewed every 1 to 3 years.
- Develop radar **workshops and seminars** for traffic adjudication personnel.
- Establish **State-level policy / procedural guidelines** to ensure proper use.

Table 3.10-6 -- **DOT Recommendations**

NHTSA recommended standards based on the 1980 test may be found under the title: Performance Standards for Speed Measuring Devices, United States Department of Transportation and National Highway Traffic Safety Administration, Federal Register, Volume 46, Number 5, January 8, 1981. Below is a list of some of the recommended specifications.

| Performance Characteristics | Minimum Requirement |
|---|---|
| **Frequency** | |
| X Band | 10.525 GHz ±   25 MHz |
| K Band | 24.150 GHz ± 100 MHz |
| **Beamwidth** | |
| X Band | 18° |
| K Band | 15° |
| **Minimum Detection Range** | 500 feet |
| **Accuracy** | |
| Stationary Mode | ± 1 mph |
| Moving Mode | ± 2 mph |
| **Tuning Fork Tolerance** | 0.5% |

Table 3.10-7 -- **DOT Performance Recommendations**

NOTES

# Chapter 4 -- Across-the-Road Radar

## Chapter 4.1 -- Photo Radar

### Operation

Photo radars are *Across-the-Road* radars, the beam is pointed across the road instead of *Down-the-Road*. Across the road radar angles the beam across the road at 20° to 23°.

The antenna must be located such that traffic is within the minimum and maximum design limits. Too close and traffic passes through the beam too fast, too far and reflected signals may be too weak or processing algorithm errors occur. The camera must be aligned for approaching or receding traffic.

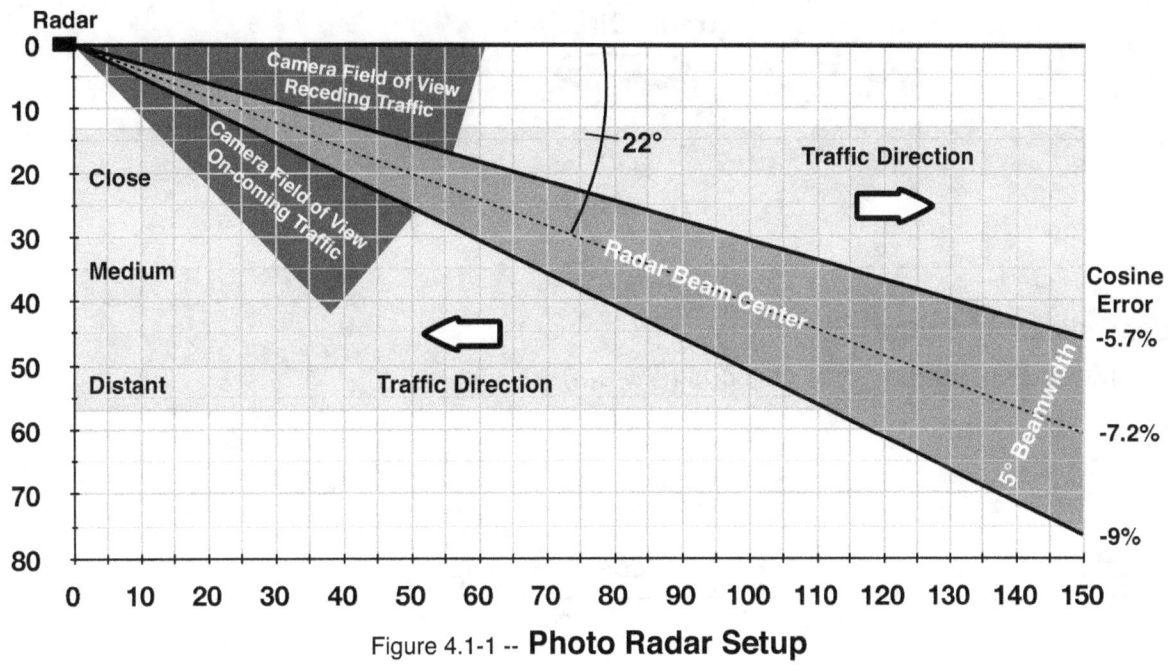

Figure 4.1-1 -- **Photo Radar Setup**

| Typical Setup Requirements | |
|---|---|
| Antenna Distance From Lanes: | Minimum Distance 18 feet<br>Maximum Distance 50 feet |
| Antenna Height above Ground: | 3 - 6 feet relative to traffic lanes |
| Camera Alignment: | Set to Photograph approaching or receding Traffic |

## Alignment Speed Errors

Antenna alignment effects accuracy, pointing into traffic, greater align angle, causes low speed readings and pointing away from traffic, smaller angle, causes high speed readings.

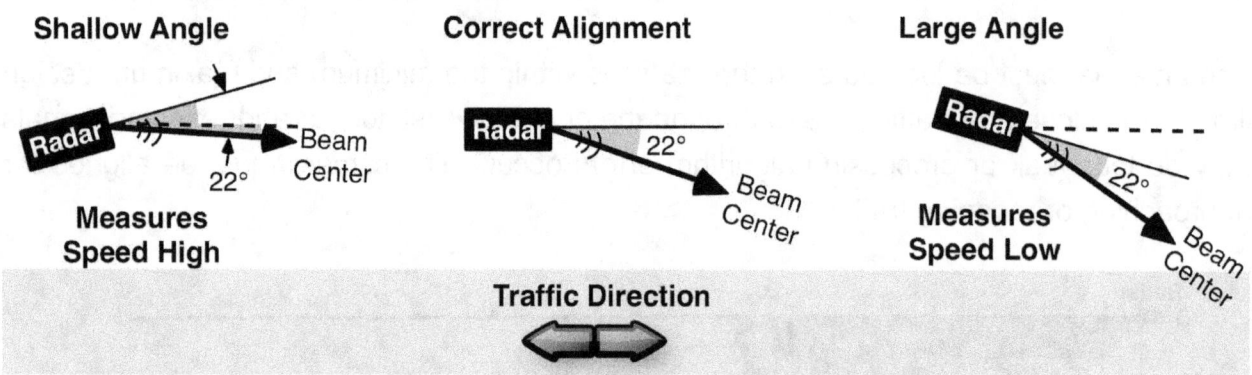

Figure 4.1-2 -- **Alignment Error**

Radar Calculated Speed   $v_c = v_m / cos\ \beta$
Radar Measured Speed     $v_m = v_o\ cos\ (\beta \pm \beta_{err})$

### Radar Calculated Speed

$$v_c = v\ \frac{cos\ \beta + cos\ \beta_{err}}{cos\ \beta}$$

$\beta$ = alignment angle              $v_c$ = radar calculated speed
$\pm\ \beta_{err}$ = alignment angle error       $v$ = traffic speed

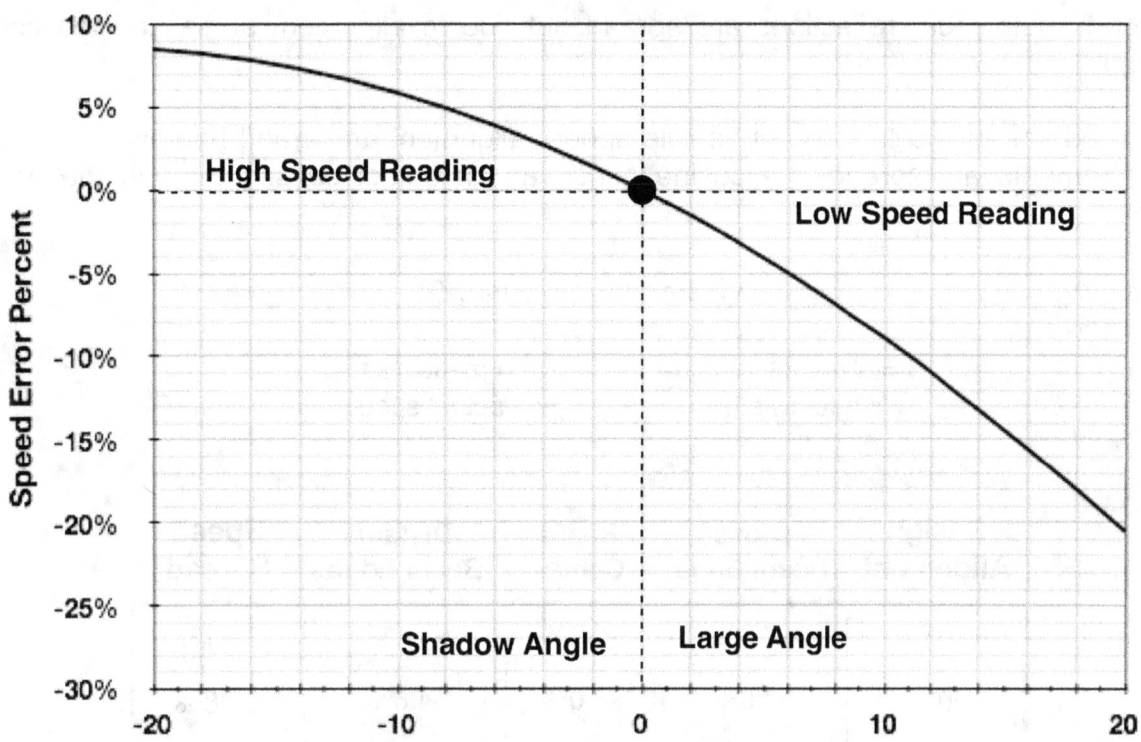

Figure 4.1-3 -- **Alignment and Speed Error**

## Speed Measurement Uncertainty

Vehicle reflections are not nice clean narrow signals as with down-the-road radar. Across-the-road photo radar reflections are spread out due to alignment angle and the cosine effect.

Measured speed is a function of vehicle speed, alignment angle and beamwidth. Speed measurements are spread, the cosine effect, as an object passes through the angled beam.

$$v_m = v \cos (ß \pm ß_w / 2)$$

ß = alignment angle      $v_m$ = measured speed
$ß_w$ = beamwidth      v = actual speed

| Design Alignment | Close Beam Edge | Beam Center | Distant Beam Edge | Speed Spread |
|---|---|---|---|---|
| 19.5° | -7.3% | -5.7% | -4.4% | 2.91% |
| 20.0° | -7.6% | -6.0% | -4.6% | 2.98% |
| 20.5° | -7.9% | -6.3% | -4.9% | 3.06% |
| 21.0° | -8.3% | -6.6% | -5.2% | 3.13% |
| 21.5° | -8.6% | -7.0% | -5.4% | 3.20% |
| 22.0° | -9.0% | -7.3% | -5.7% | 3.27% |
| 22.5° | -9.4% | -7.6% | -6.0% | 3.34% |
| 23.0° | -9.7% | -7.9% | -6.3% | 3.41% |
| 23.5° | -10.1% | -8.3% | -6.6% | 3.41% |
| 24.0° | -10.5% | -8.6% | -7.0% | 3.55% |

Table 4.1-1 -- **Speed Error and Spread for a 5° Beamwidth**

In reality the reflected spectrum has a greater spread than can be accounted for by the cosine effect and vehicle speed. Empirical data was taken using a radar with a 5° beamwidth angled 20° across the road against a 20 mph vehicle. The cosine effect predicts a spread of 18 - 19 mph, however the data has a spread of 13 - 21 mph. The difference is due to target vehicle *wheel rotation* adding Doppler reflections.

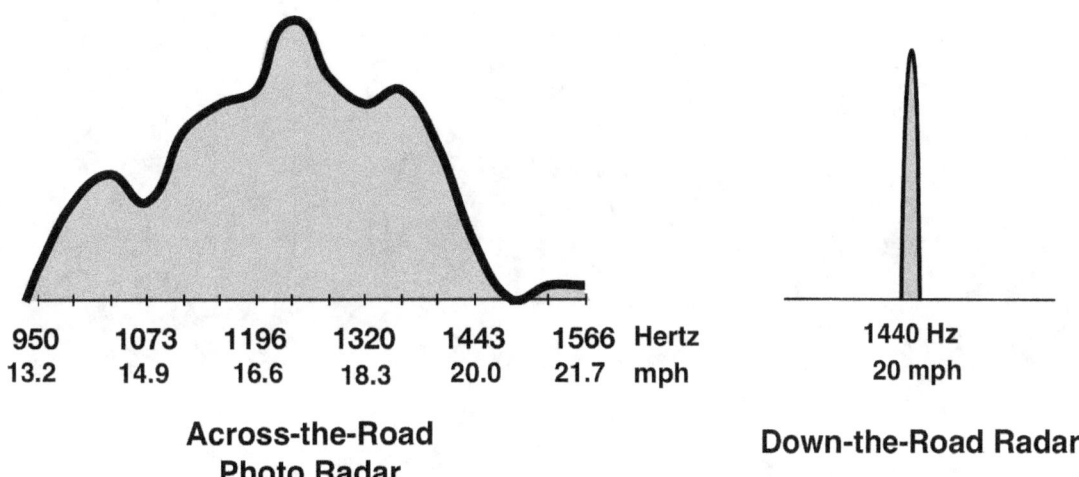

| 950 | 1073 | 1196 | 1320 | 1443 | 1566 | Hertz |
| 13.2 | 14.9 | 16.6 | 18.3 | 20.0 | 21.7 | mph |

**Across-the-Road
Photo Radar**

1440 Hz
20 mph

**Down-the-Road Radar**

Figure 4.1-4 --
**Across-the-Road Doppler
Spectrum / Speed Spread**

K Band Radar
20° Alignment, 5° Beamwidth
20 mph Target Vehicle
*Calibration Techniques for Across-the-Road Traffic Radars*
National Institute of Standards and Technology, NIST Technical Note 1398, May 1998.

Manufacturers are reluctant to discuss how their radar processes the speed scatter, claim as proprietary information and trade secret. Across the road radars have fractions to a couple of seconds to measure speed. The Cosine Effect error and wheels rotating produce reflections that are not steady stable signals. Across-the-road photo radars are less accurate and less reliable than down-the-road radars.

NOTES

# Chapter 5 -- Laser Radar (Lidar)

## Chapter 5.1 -- Introduction to Laser Radar

Laser radars, also referred to as lidars, use pulsed infrared laser light invisible to the eye instead of continuous microwaves.

### Acronyms

LASER   - **L**ight **A**mplification by **S**imulated **E**missions of **R**adiation

LIDAR    - **L**ight **D**etection **A**nd **R**anging

LADAR   - **LA**ser **D**etection **A**nd **R**anging

Laser radars transmit pulsed infrared laser light to measure target range. The time it takes a pulse to travel at the speed of light from the lidar to the target and back is used to compute range. The change in range over time is used to calculate speed. Laser radars typically require 0.3 to 0.7 seconds *sample time* to get one speed reading. Tens to hundreds of pulses are used to calculate one speed reading.

Laser radars operate from a stationary position only, no moving mode, and measure speed of approaching or receding traffic. Most models also have a range only mode for measuring the range of stationary objects.

The Federal Communications Commission (FCC) regulates radiated emissions from high speed circuits such as the processing circuits inside a lidar, but not light frequencies or wavelengths. The Federal Drug Administration (FDA) Center for Devices and Radiological Health (CDRH) regulates laser products sold in the United States. Police laser radars are Class 1 devices, by American National Standards Institute (ANSI) definition, and considered eye-safe based on current medical knowledge. Even so do **not** stare or look at a laser aperture while transmitting, especially on beam.

**Infrared Spectrum**

Laser radars radiate in the infrared (IR) part of the spectrum at 904 nanometers wavelength. Visible light is between 700 nanometers (red) and 400 nanometers (violet).

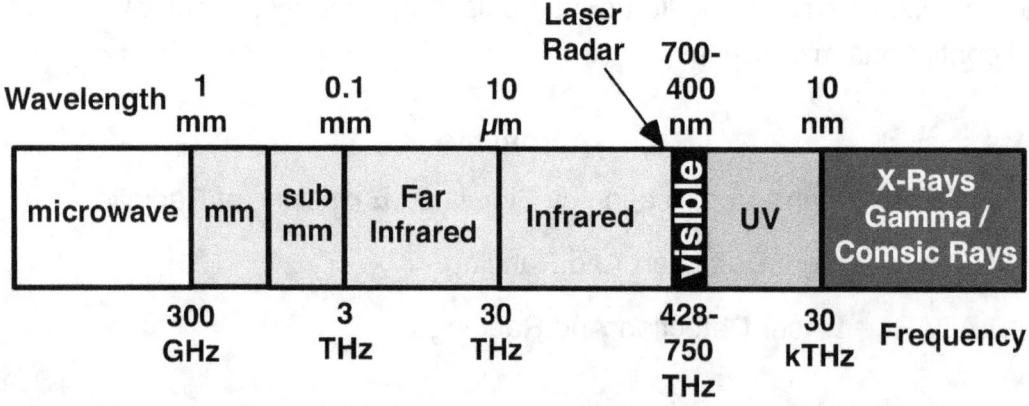

Figure 5.1-1 -- **Upper Spectrum**

mm = millimeter　　　μm = micrometer　　　nm = nanometer

Light is measured in wavelength instead of frequency. Common wavelength dimensions include angstroms, nanometers, and microns.

| unit | Symbol | meters | inches |
|------|--------|--------|--------|
| angstrom | Å | $10^{-10}$ | 0.000 000 004 |
| nanometer | nm | $10^{-9}$ | 0.000 000 039 |
| micrometer or micron | μm | $10^{-6}$ | 0.000 000 370 |

**Laser Radar Wavelength**

9,040 Å = 904 nm = 0.904 μm

# Range and Speed Measurements

Laser radars are really laser range finders adapted to calculate speed by measuring range difference over a period of time. These instruments are pulse modulated and transmit narrow infrared laser light pulses at a fixed rate. Range is determined by measuring the time it takes a single pulse to travel from the lidar to the target vehicle and back. At least two pulses are required to calculate speed. In practice 10's to 100's of pulses are processed to determine one speed measurement.

Pulse round trip time is measured to determine range. Target vehicle range is half the round trip distance the pulse travels, and half the time.

$$R = c\,t\,/\,2$$

R = Range from Lidar to Target Vehicle
t = Pulse Round Trip Travel Time
c = Speed of Light

## Data Set Collection

The lidar collects a series of range measurements during a sample time period before calculating speed. To distinguish legitimate echoes from false returns each echo pulse must be within expected pulse widths limits to be used as a valid data point. An entire data set must meet variance and standard deviation constraints to be considered valid. A valid data set is usually processed with a linear least squares line algorithm to line fit the change in range for computing speed.

Early lidars discarded an entire data set if only one echo was missing or out of place. Modern lidars process speed if a limited number of echoes are missing or out of expected bounds. Manufacturers do not disclose, claim proprietary, any details on the criteria used to determine valid data sets making independent scrutiny impossible.

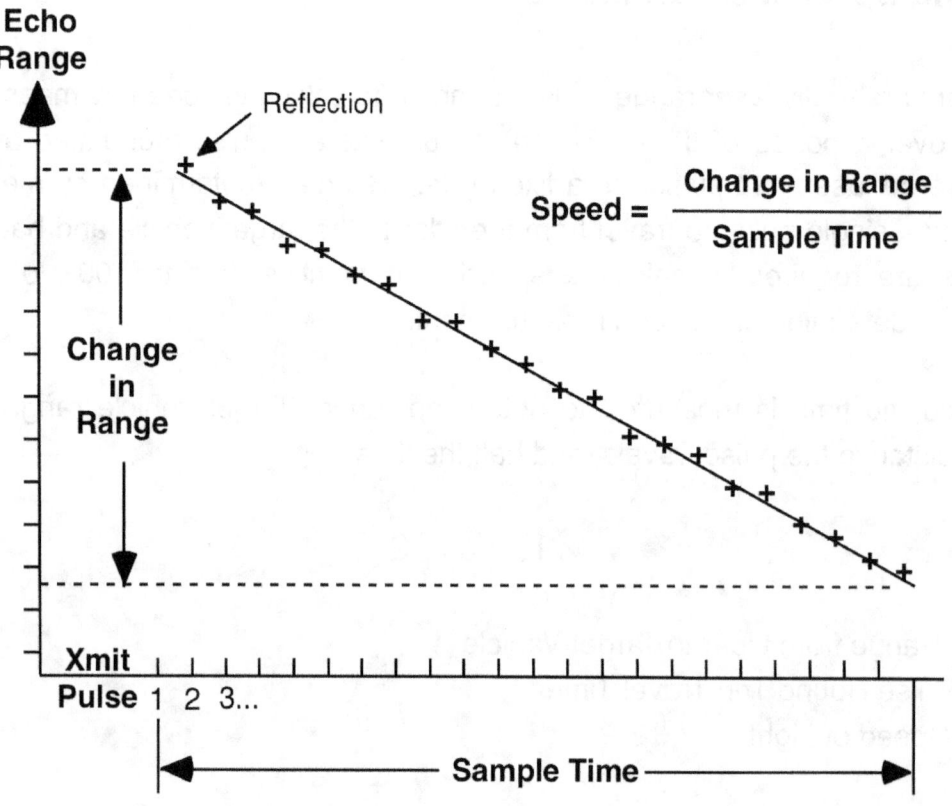

$$Speed = \frac{Change\ in\ Range}{Sample\ Time}$$

Figure 5.1-2 -- **Lidar Data Set**

## Speed Calculation

After a data set is determined valid the lidar calculates velocity from change in range divided by sample time. Decreasing ranges indicate the vehicle is moving toward the lidar, increasing ranges the vehicle is moving away. Under ideal conditions the lidar could display a speed reading after only one sample period. In practice a second or more is more realistic. Bad conditions, fog, rain, carbon dioxide, obstructions, could force the lidar to take as long as several seconds or miss a target entirely.

## Laser Systems

Laser systems use diodes to generate a laser pulse for transmission to an optical aperture. A diode detector sensitive to IR converts the reflections to electrical signals. A computer processes the reflection's amplitude, pulse width, and time of arrival to determine if the reflection is legitimate.

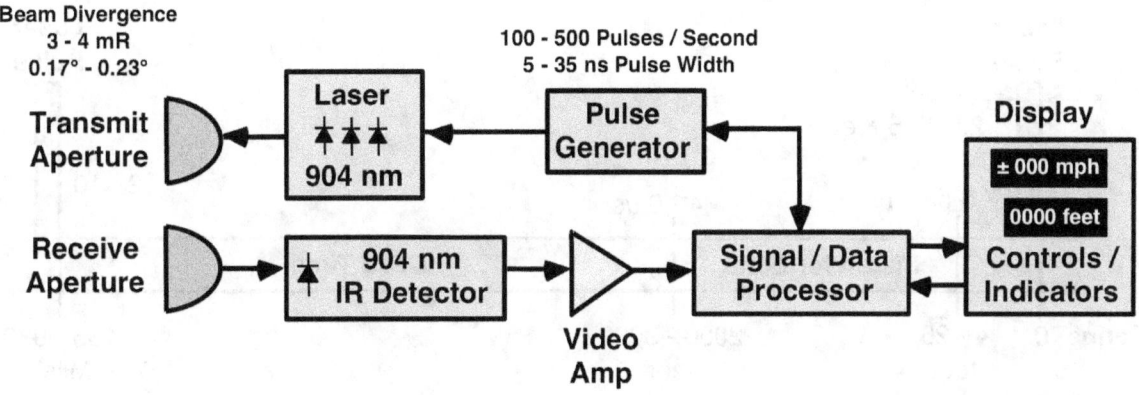

Figure 5.1-3 -- **Lidar Block Diagram**

Laser radar apertures, light antennas, are optical focusing devices, lenses, prisms, and mirrors, used to collimate laser energy into a narrow beam. Some models use the same aperture for transmit and receive, some use separate apertures. Some multiple apertures are side-by-side, others are one on top of the other where the top aperture radiates.

Laser radars typically use 3 semiconductor diodes to generate infrared laser light. Return signals, reflections, are routed directly from the aperture to an infrared detector. The detector converts infrared light to electrical energy and routes the signals to a video amplifier. The video amplifier adjusts signal gain by time, longer range signals require more gain. Video signals that exceed the detection threshold get processed for leading edge detection by range gates for tracking. If the data set passes the preprogrammed test criteria target data can then be calculated and displayed.

A programmable processor allows the flexibility to program different pulse widths, PRF's (Pulse Repetition Frequency) and algorithms for computing speed. Exact parameters vary with manufacturer, model, and even software version. Pulse transmit rates are between about 100 to 500 pulses per second. Sample times will be on the order of 0.3 to 0.7 seconds. Pulse widths have been reported to be as low as 5 ns and as high as 35 ns.

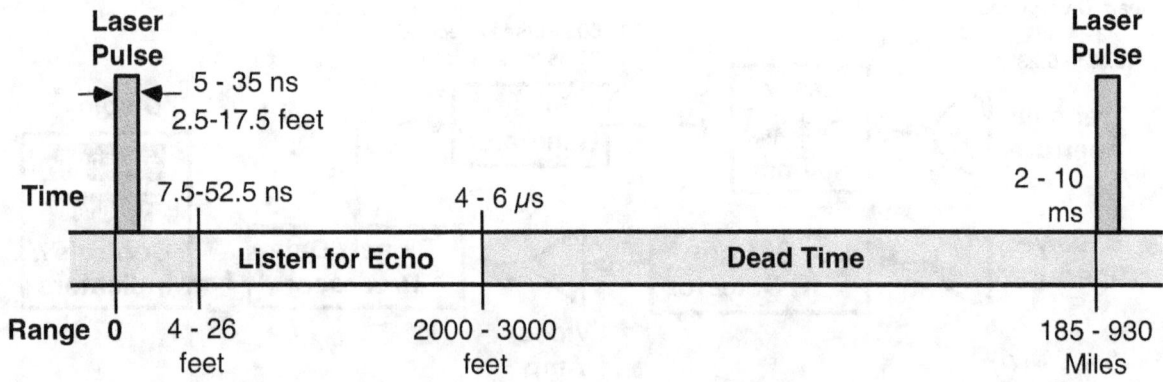

Figure 5.1-4 -- **Lidar Waveform**

Minimum range is a function of pulse width. To receive an entire pulse, as opposed to a fraction, target range must be at least half the pulse width in range for a total of 1.5 pulse widths from time zero. Pulse width minimum range varies from about 4 to 26 feet.

### Lidar Specifications

| | |
|---|---|
| Band: | Infrared |
| Wavelength: | 904 nm ± 5 nm |
| Beam Divergence: | 3 - 4 milliradians (0.17° - 0.23°) |
| Sample Time: | 0.3 - 0.7 seconds |
| Maximum Range: | 2000 - 3000 feet |
| PRF: | 100 - 500 pulses / second |
| Pulse Width: | 5 - 35 nanoseconds |

# Chapter 5.2 -- Lidar Operations

## Physical Configurations

Laser radars are single unit hand held devices.  Some can be mounted to a tripod to make it easier to keep the narrow beam on target.   Some models have an optional shoulder mount.  Some units are designed like oversized binoculars.

|  | Hand Held | Binocular Form |
|---|---|---|
| **Aiming** | Heads Up Display or Scope | Cross-hairs Projected onto Eyepiece. |
| **Display** | Large Speed and Range Display | Speed and Range Projected onto Eyepiece. |
| **Power Source** | Internal Batteries and/or Vehicle Battery DC plug. | Internal Battery |
| **Mounting Options** | • Tripod <br> • Shoulder Mount | |
| **Timing Mode** | Most Models | |
| **Laser Mode** | Stationary ||
| **Range Only** | Most Models ||
| **Traffic** | On-coming (+) and/or Going (-) ||
| **Transmissions** | Instant On ||
| **Speed Tone** | Audio Tone Proportional to Speed ||

Table 5.2-1 -- **Lidar Configurations**

| Status | Mode |
|--------|------|
| XMIT | Transmitting |
| STBY | Standby |
| RFI | Radio Frequency Interference |
| BIT Err | Built-In-Test Error |
| Low Volt | Low Voltage condition |

Table 5.2-2 -- **Status and Fault Indicators**

Most lidar manufacturers caution not to point the aperture directly at the sun. Direct sunlight has enough energy to cause damage by burning out the infrared detector diodes.

Many lidars have a Timing Mode. The range between two points is measured and entered into the lidar. The operator pushes a button when a target passes between the two points and the lidar displays the calculated speed without ever transmitting.

## Beamwidth

Laser radars have narrow beams on the order of 3 - 4 milliradians (mR), 0.17° - 0.23°. The narrow beam allows the operator to single out vehicles at relatively short ranges. Lidars also require the operator to precisely aim the laser at the same point on the target vehicle.

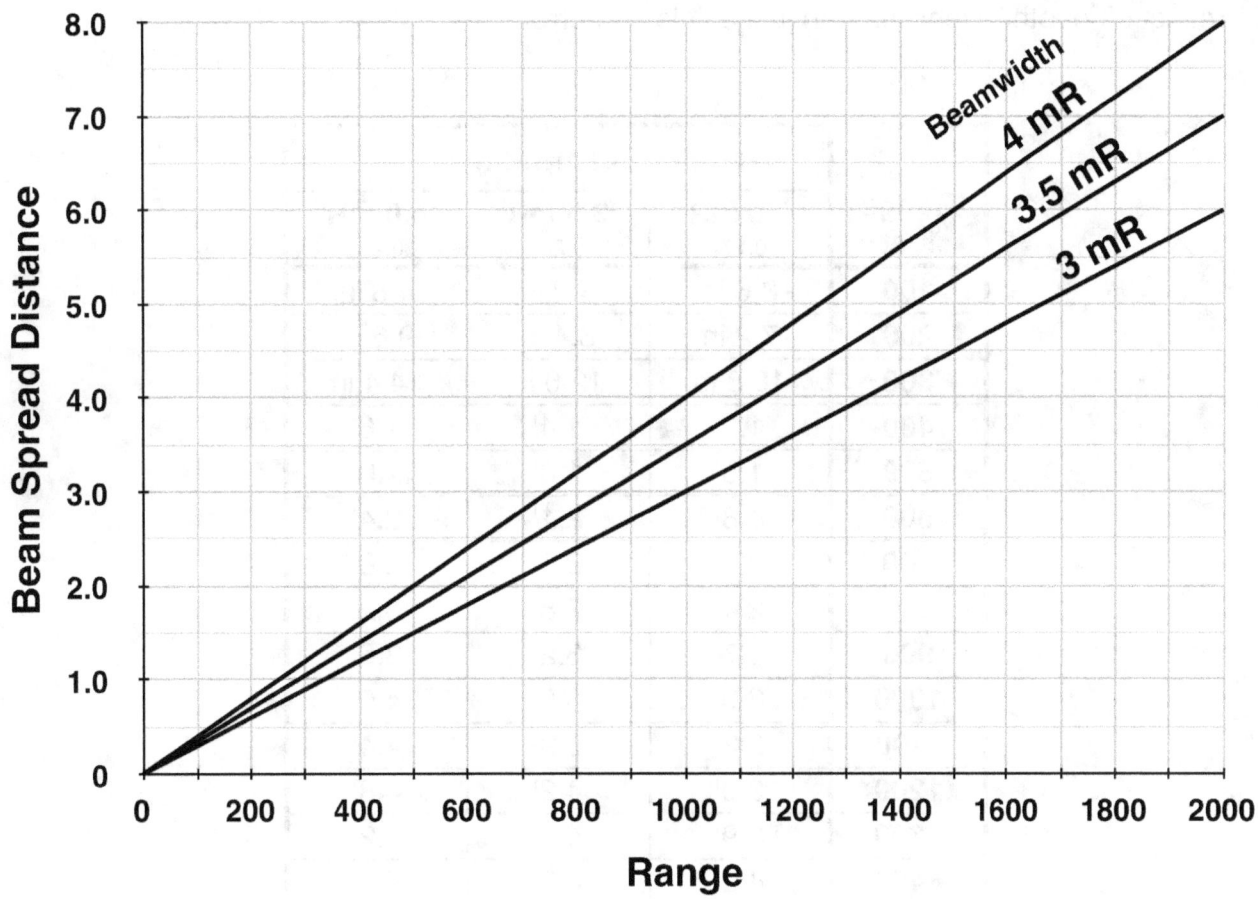

Figure 5.2-1 -- **Laser Radar Beam Spread**

$$d = 2 R \tan (\beta / 2)$$

d = beam spread distance       ß = beamwidth or beam divergence
R = range

In the figure and equation above, range and beam spread distance are in the same unit dimensions, feet, meters, etc.

Laser radars are very selective at ranges less than 500 feet, beam spread is a couple of feet or less.  Over 500 feet the beam can reflect off of different parts, different ranges, of a target vehicle introducing errors.  At 1000 feet the beam spread is 3 - 4 feet, wide enough to reflect off of other vehicles with just a little aim error.

| Range feet | Beamwidth | | |
|---|---|---|---|
| | 3.0 mR 0.17° | 3.5 mR 0.20° | 4.0 mR 0.23° |
| 100 | 3.6 in | 4.2 in | 4.8 in |
| 200 | 7.2 in | 8.4 in | 9.6 in |
| 300 | 10.8 in | 12.6 in | 14.4 in |
| 400 | 1.2' | 1.4' | 1.6' |
| 500 | 1.5' | 1.8' | 2.0' |
| 600 | 1.8' | 2.1' | 2.4' |
| 700 | 2.1' | 2.5' | 2.8' |
| 800 | 2.4' | 2.8' | 3.2' |
| 900 | 2.7' | 3.2' | 3.6' |
| 1000 | 3.0' | 3.5' | 4.0' |
| 1100 | 3.3' | 3.9' | 4.4' |
| 1200 | 3.6' | 4.2' | 4.8' |
| 1300 | 3.9' | 4.6' | 5.2' |
| 1400 | 4.2' | 4.9' | 5.6' |
| 1500 | 4.0' | 5.3' | 6.0' |
| 1600 | 4.8' | 5.6' | 6.4' |
| 1700 | 5.1' | 6.0' | 6.8' |
| 1800 | 5.4' | 6.3' | 7.2' |
| 1900 | 5.7' | 6.7' | 7.6' |
| 2000 | 6.0' | 7.0' | 8.0' |

Table 5.2-3 -- **Beam Spread in feet (') or inches (in)**

**Detection Range**

Some lidar manuals suggest aiming the beam at the license plate for best performance. Maximum detection range is maybe a half mile in good weather. A vehicle can be made stealthy by removing the license plate and covering all chrome parts, headlights, tail lights, parking lights, and turn signal lights with black tape.

Atmospheric Conditions

Atmospheric conditions effect propagation and detection range. The best conditions for detection is an atmosphere that is cool, dry, and clear. Fog, rain, smoke, dust particles, carbon dioxide and water vapor reduce detection capability, sometimes to zero.

### Lidar Range in Weather

| Condition | Range Reduction |
|---|---|
| Fog | Most Severe |
| Rain Downpour | Severe |
| Heavy Rain | Severe |
| Rain / Drizzle | Noticeable |
| Dust Particles | Noticeable |
| Atmospheric Gases<br>• Carbon Dioxide ($CO_2$)<br>• Water Vapor ($H_2O$) | Slight |

Snow and sleet reduce detection much like rain, the more water molecules the greater the reduction.

## Lidar Test and Calibration

Laser radars measure small range changes in short periods of time, this imposes tight tolerances on internal timing circuits.  Aperture(s) to aim spot alignment is critical and also has very tight tolerances.  The units should be tested by the operator before use.

### Operator Testing

| Static Test and Checks | | |
|---|---|---|
| 1 | Check Lidar Calibrated. | Most states require a lidar be tested by a certified shop periodically, typically once or twice a year.<br><br>A sticker on the lidar, or records, should indicate last calibration test, next (due) test, and who tested. |
| 2 | Run Lidar Self-Test. | Self test should be run before, during, and after use. |
| 3 | Check Beam Alignment. | Lidar is swept pass a utility pole or a stop sign at 250 - 500 feet or greater. A range reading should only occur when object sweeps across the aim circle or crosshairs. Vertical and horizontal angles should be checked. |
| 4 | Test Stationary Object Range. | Lidar range measurement checked against a utility pole or stop sign of know distance, 250 - 500 feet or greater. |
| Operational Test | | |
| 5 | Test Lidar against Vehicle of Known Speed. | Test vehicle should have a calibrated speedometer.  Best to run test at operational location. |

Tuning forks do not register speed on laser radar.

External optical surfaces, aperture, aim scope or heads up display, should be periodically cleaned exactly as recommended by the manufacturer.  Additionally all optical surfaces should be covered or capped when not in use to prevent damage.  Scratches, pits, and stains in the aperture can degrade detection performance, beam direction and alignment to aiming device.

## Lidar Self-Test

### Self Test runs;
- on power up
- operator initiated
- automatically after a speed is locked
- any or all the above

### Test / Automatic Adjustments
- supply voltage - car or internal battery
- a test target signal
- portions of the digital circuits
- optics
- display indicators
- any or all the above

## Laboratory / Shop Calibration

Some agencies require periodic, once or twice per year, additional checks and tests above and beyond everyday routine testing.  Rigorous tests should include;

- speed accuracy for multiple speeds
- range accuracy for multiple ranges
- transmit power to radiation aperture
- radiation field strength at 3 meters or more
- vertical and horizontal beam divergence
- IR detector sensitivity
- power supply limits (± volts)

# Chapter 5.3 -- Lidar Operational Problems

## Acceleration Limits

Laser radars, like microwave radars, are designed to measure vehicles traveling at a relatively constant speed. Vehicles changing speed greater than the accuracy during one sample period cannot be measured, speed is changing too fast. Laser radars typically have longer sample times making the acceleration limits more narrow.

### Acceleration Limit = accuracy / sample period

$$a_{max} = \pm\, v_{acc} / t_i$$

$a_{max}$ = Maximum Acceleration     $\pm v_{acc}$ = Speed Accuracy     $t_i$ = sample time

Common sample periods, varies with model, 300, 333, 400, 500, 700 milliseconds - 0.25, 0.3, and 1/3 seconds.

Table 5.3-1 -- **Acceleration Limits**

| Sample Period | | Maximum Change in Speed |
|---|---|---|
| seconds | milliseconds | Miles per Hour per Second |
| 0.3 | 300 ms | ± 3.3 mph / second |
| 1/3 | 333 ms | ± 3.0 mph / second |
| 1/2 | 500 ms | ± 2.0 mph / second |
| 0.7 | 700 ms | ± 1.4 mph / second |
| 1 | 1000 ms | ± 1.0 mph / second |

## Sweeping or Scanning Error

Scanning or sweeping the lidar beam across the ground such that range increases, or decreases, in a steady manner will produce a speed reading. The measured speed is a function of the change in range over sweep time. Increasing range short to long produces a receding speed, sweeping long to short range produces an approaching speed.

$$v_m = R_d / (n\ t_i)$$

$$R_d = v_m\ n\ t_i$$

$v_m$ = measured speed           $t_i$ = lidar sample time
$R_d$ = distance lidar scanned    $n$ = number of sample periods

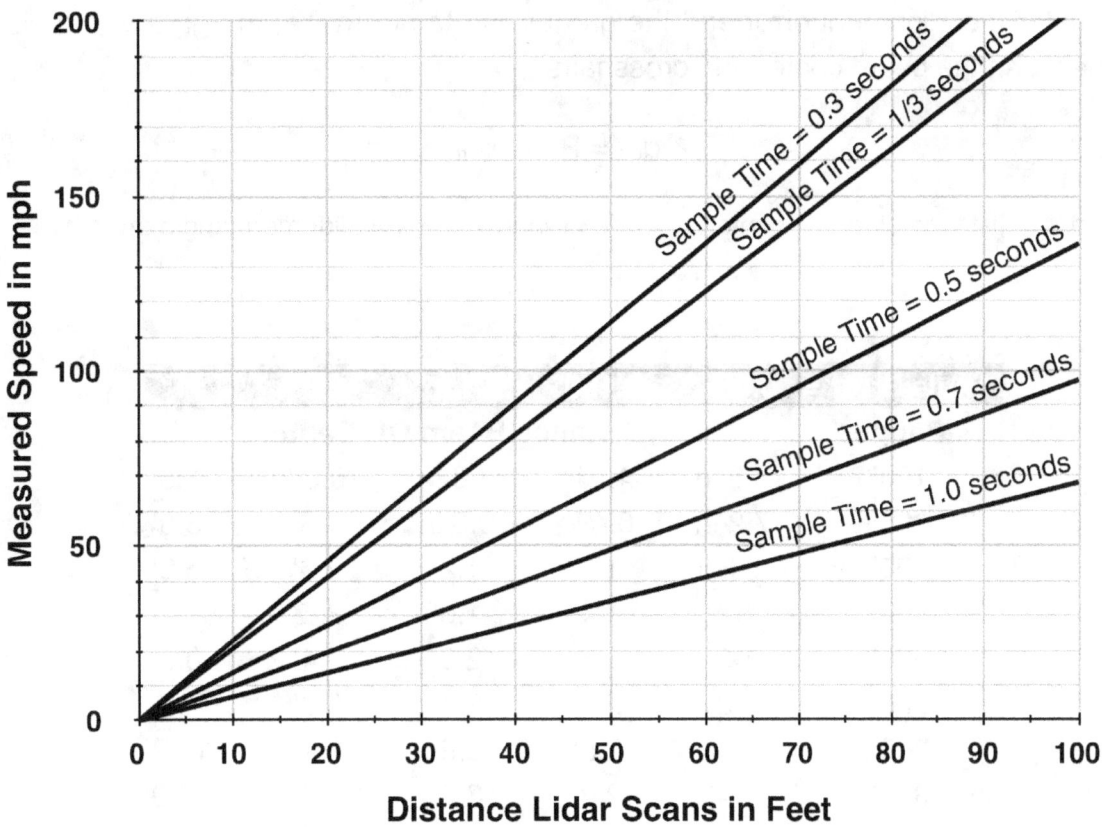

Figure 5.3-1 -- **Range versus Speed for 1 Sample Period**

If speed in mph and range in feet;

$$v_m = (15/22)\ R_d / (n\ t_i)$$

$$R_d = 22\ v_m\ n\ t_i / 15$$

## Aiming and Alignment Errors

The narrow laser beam and aim crosshairs must be closely aligned to very tight tolerances. Any misalignment introduces aiming errors. For the crosshairs to fall in the beam the alignment error must be less than half of the beamwidth. An alignment error that is greater than half the beamwidth will place the crosshairs outside the beam.

| | | | |
|---|---|---|---|
| Beam Divergence: | 3.0 mR | 3.5 mR | 4.0 mR |
| | 0.17° | 0.20° | 0.23° |
| Half Beamwidth: | 1.5 mR | 1.25 mR | 2.0 mR |
| | 0.086° | 0.100° | 0.115° |

The greater the alignment error and the greater the target range, the greater the distance error between the beam center and crosshairs.

$$d_{err} = R \ tan \ \beta_{err}$$

$d_{err}$ = distance off beam center        R = range        $\beta_{err}$ = alignment angle error

| Angle Error = | 0.086° | 0.100° | 0.115° | 0.5° | 1° |
|---|---|---|---|---|---|
| Range | Distance Beam Off Center | | | | |
| 100' | 3.6 in | 4.2 in | 4.8 in | 10.5 in | 1.8' |
| 200' | 7.2 in | 8.4 in | 9.6 in | 1.8' | 3.5' |
| 300' | 10.8 in | 1.1' | 1.2' | 2.6' | 5.2' |
| 400' | 1.2' | 1.4' | 1.6' | 3.5' | 7.0' |
| 500' | 1.5' | 1.8' | 2.0' | 4.4' | 8.7' |
| 600' | 1.8' | 2.1' | 2.4' | 5.2' | 10.5' |
| 700' | 2.1' | 2.5' | 2.8' | 6.1' | 12.2' |
| 800' | 2.4' | 2.8' | 3.2' | 7.0' | 14.0' |
| 900' | 2.7' | 3.2' | 3.6' | 7.9' | 15.7' |
| 1000' | 3.0' | 3.5' | 4.0' | 8.7' | 17.5' |

Table 5.3-2 -- **Distance in feet (') or inches (in) Off Beam Center given Range and Alignment Angle Error**

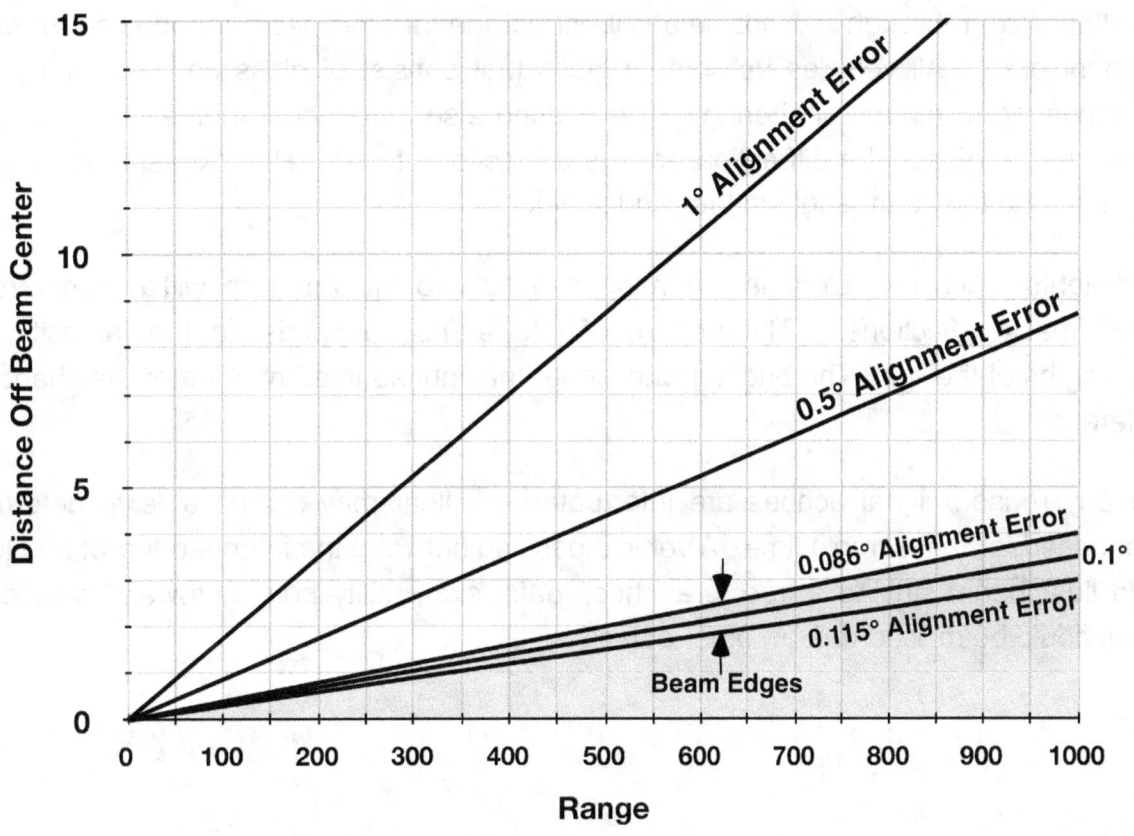

Figure 5.3-2 -- **Distance Off Beam Center versus Range
for Alignment Angle Errors**

A 3 milliradian beamwidth edge occurs 0.086° off center, a 4 milliradian beamwidth edge occurs 0.115° from beam center.

The operator must aim the beam to an accuracy of within tenths of a degree, the aperture must be aligned within hundredths of a degree to the optical aiming device. A two-aperture system must align both apertures to thousandths of a degree, and this combination must then be aligned to the optical aiming device.

## Windshield and other Interference

Operating a lidar through a windshield will diffract the infrared beam introducing an angle alignment error.  All vehicles use safety glass that consist of glass covered with a thin plastic coating to reduce shattering.  The coating also introduces another angle error in addition to the glass. The diffraction error is a function of material thickness and index of refraction, and the beam angle to the windshield.

Bright lights, such as halogens, beaming directly into the aperture will desensitize or entirely mask reflections.  The degree of interference depends on the intensity and wavelengths of the light. The brighter and closer the light source, the greater the chance of interference.

If for any reason signal echoes are interrupted the lidar may not be able to determine target speed for that sample time.  A vehicle passing between the intended target and lidar, or the lidar beam striking a tree, branches, leafs, sign, utility pole or tower, some or all returns could be missed or from other objects.

## Speed Error Due To Range Errors

### Reflection from Different Vehicle Parts and Sections

Laser radars are designed to track flat surfaces on a target vehicle.  Manufactures suggest aiming at the license plate.  Reflections that do not come from the same section, range, of the vehicle will cause range errors.  Tracking from the back to front of a vehicle causes a **high** measured speed.   Tracking from the front to back of a vehicle causes a **low** measured speed.

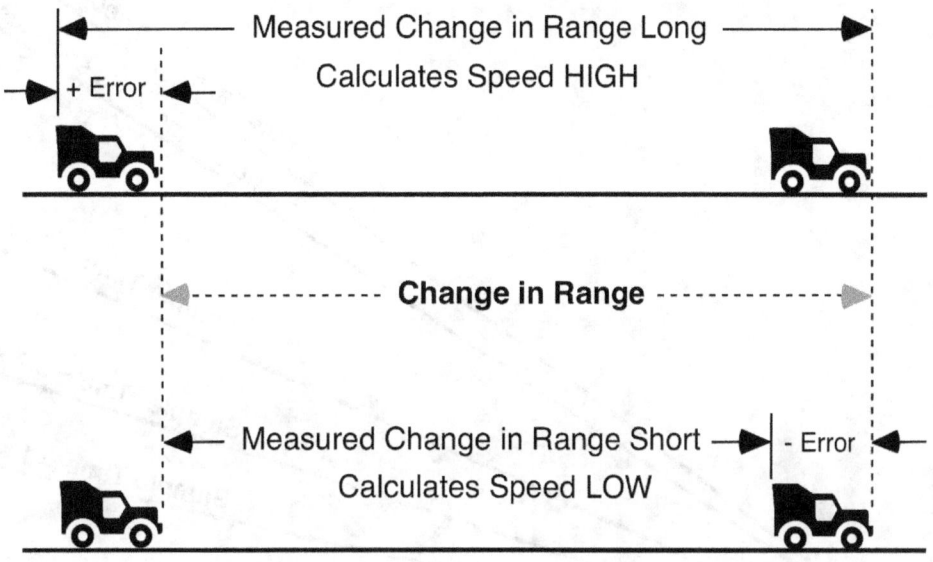

Figure 5.3-3 -- **Range and Speed Error**

Measured speed given range and distance error, and sample time.

$$v_m = ( R \pm d_{err} ) / t_i$$

$v_m$ = measured speed          $d_{err}$ = distance error (vehicle length)
$R$ = actual change in range      $t_i$ = sample time

Speed error for range error and sample time.

$$v_{err} = \pm d_{err} / t_i$$

$v_{err}$ = speed error
$d_{err}$ = distance error
$t_i$ = sample time

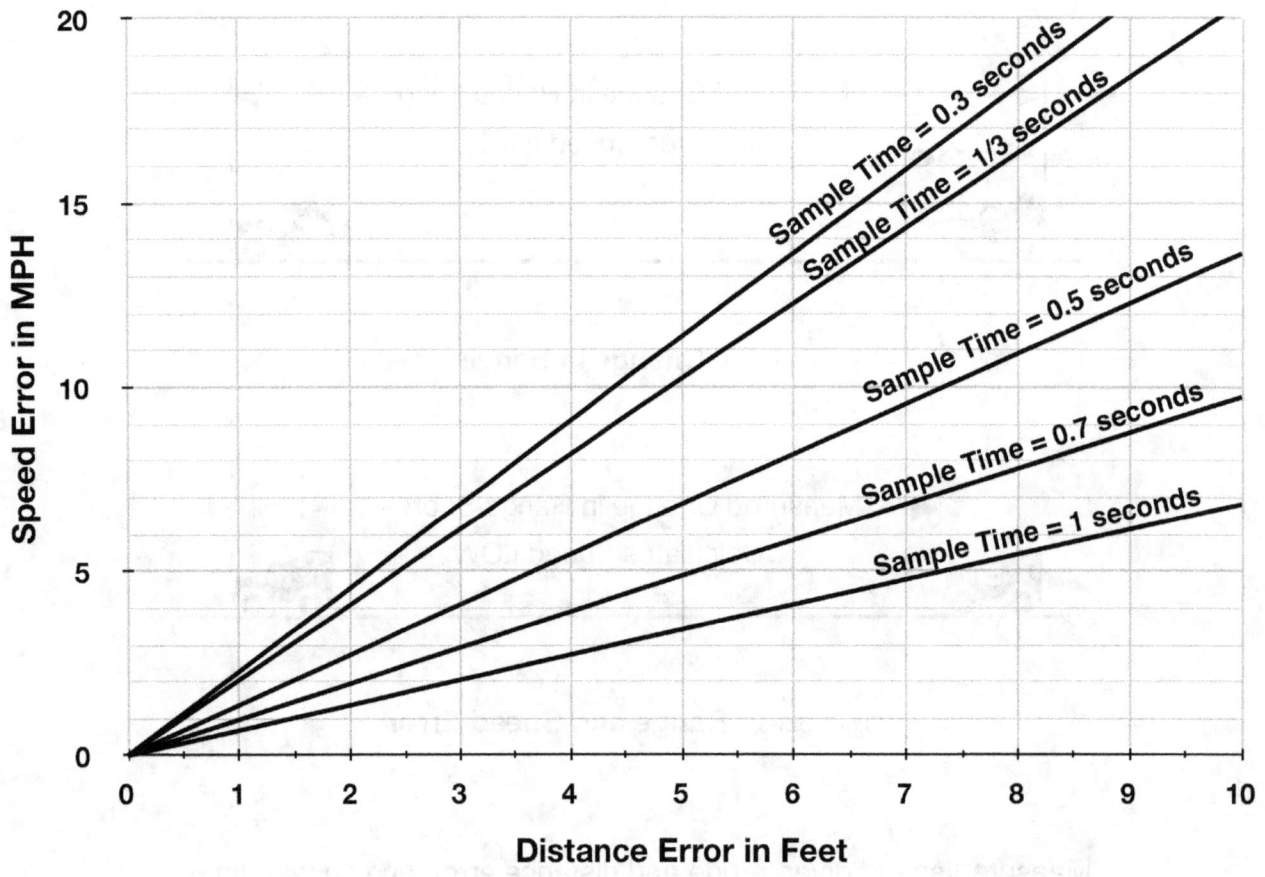

Figure 5.3-4 -- **Distance Error versus Speed Error**

## Range Resolution

Lidar pulse width determines range resolution, the ability to separate closely spaced objects in the range dimension.  The more narrow the pulse the higher the resolution.  A high resolution narrow pulse could reflect from *different parts of a target vehicle* introducing range and speed errors.

To resolve two objects in range they must be separated by over half a pulse width. Pulse widths vary from 5 to 35 nanoseconds.

$$d_s > c \, t_{pw} / 2$$

$d_s$ = object separation
$t_{pw}$ = lidar pulse width
$c$ = speed of light

$$d_s > 0.5 \, t_{pw}$$

$d_s$ = object separation in feet
$t_{pw}$ = pulse width in nanoseconds

| Pulse Width nanoseconds | Object Separation |
|:---:|:---:|
| 5 ns | 2.5 feet |
| 10 ns | 5.0 feet |
| 15 ns | 7.5 feet |
| 20 ns | 10.0 feet |
| 25 ns | 12.5 feet |
| 30 ns | 15.0 feet |
| 35 ns | 17.5 feet |

Table 5.3-3 -- **Distance Between Objects**

NOTES

# Chapter 6 -- Before / After a Speeding Ticket

## Chapter 6.1 -- Motoring  Tips

**Driving Tips**

If one knows of or is warned from a CB or a passing motorist flashing headlights of an impending speed trap it's a good idea to drop speed five mph below the limit.  This may or may not prevent a ticket, but could help in court if a ticket is issued.  If one can claim the vehicle speed was at least five mph below the limit speedometer accuracy will be less questionable.

If a driver believes such as a warning from a radar detector or a visual sighting a radar is operating a good countermeasure is to gently decelerate at a rate of about 3 or 4 mph per second.   A microwave or laser radar will experience problems measuring vehicles decelerating, or accelerating.

Some police switch on the radar only when ready to target a vehicle.   Some radars transmit only long enough to get a speed reading.  Depending on the situation a driver may only have a fraction of a second to respond, and in some cases may not be warned at all until after the radar gets a speed reading.  The lesson - do not depend solely on a radar detector to warn a radar is operating in the area.

Laser radars have a narrow beam making it difficult to detect unless the laser is aimed directly at the laser detector.  At ranges greater than 500 feet or more a detector may catch the wider beam spread.  A laser detector may sense a signal reflecting off another vehicle or a stationary object, but this condition usually has a short duration if it occurs at all.  Laser radar detectors are not nearly as effected as microwave detectors.

## Traffic-Stops

If the police light you up pull over and stop at the closest safest place for you and the officer, at night try to find a well lighted area. Curves, blind spots, and mediums are poor stop locations. After you stop most law enforcement officers prefer you stay in your vehicle. When approached be polite and keep your hands visible, at night turn on your interior doom light before the officer approaches.

A typical scenario is the officer request to see a driver license, proof of insurance, and vehicle registration. Most officers ask if you know why you were stopped. Comply and answer honestly, but do not admit anything that could be used against you in court. Excuses for speeding almost never work and admit guilt.

Common Excuses* for Speed (**DO NOT USE**).

• It was unintentional.
• I was in a hurry, appointment, pick up someone, etc.
• I was being forced to speed by a tailgater.
• I think the speed limit is too slow.
• my modern car stops more quickly than cars on the road when the limit was set.
• I don't think the same limit should apply when the road is empty late at night.
• The limit doesn't apply to me because I am an above average driver.
• My speeding's acceptable because it's not much over the limit and others abuse the limit more flagrantly.

* Factor influencing driver speed choices, Department of Transport (England), Circular Roads 1/95.

Many citations require a bond, driver's license or cash and signature.   Before signing a traffic citation READ IT.  Make sure you understand the violation(s).

| CITATION AND COMPLAINT | |
| --- | --- |
| **JURISDICTION:** State, County, City | **Complaint 0000001** |
| **DRIVER:** | *Name, Address, Driver License #, Info* |
| **VEHICLE:** | *Year, Make, Model, Type, Color, Description Registration* |
| **DATE:**<br>**LOCATION:** | *Day, Month, Year, Time*<br>At or Near _____ |
| **VIOLATION:** | Speed _____ mph in a _____mph Zone<br>Other Violations _____<br>Laws / Ordinances _____ |
| **METHOD:** | __HH Radar __Radar    __Moving Radar<br>__VASCAR   __Lidar    __VASCAR<br>__Timing      __Aircraft  __other |
| **ROAD:**<br>**VISIBILITY:** | __Dry __Wet __Snow __Ice<br>__Rain __Fog __Snow __Sleet __Night |
| **TRAFFIC:**<br>**AREA:** | __Cross __Oncoming __Same Direction<br>__Rural __Bus/Ind    __Res __School<br>#____Lanes - ___Divided |
| **BONE:** | __Driver License    __none (promise)<br>__Cash __Bond   $_____ |
| **APPEARANCE:** | Location:_____<br>Date / Time:_____/_____<br>__ Appearance Required.<br>__ Appearance not Required if Guilty Plea. |

Table 6.1-1 -- **Typical Speeding Citation Form**

Any incorrect information discovered on the citation that can be proven or documented  in court may cast doubt on the creditability of the violation.

If ticketed collect as much information as possible about the circumstances. Write down *everything* you can remember as soon as possible. Besides information on the ticket some or all of the information below may be helpful when analyzing events later.

| | | |
|---|---|---|
| 1. | **Setup** | • Antenna Distance to target vehicle lane center |
| 2. | **Tracking** | • Target Vehicle Range to Radar |
| 3. | **Other Traffic** | • Near Target Vehicle<br>• Near Radar / Patrol Vehicle |
| 4. | **Potential Interference** | • Communication Towers, ground radars, etc.<br>• Power lines, transformers |
| 5. | **Radar or Lidar** | • Make and Model<br>• Frequency or Frequency Band<br>• Antenna Location and Alignment - Vehicle mounted units |
| 6. | **Patrol Vehicle** | • Dash Camera or other Video / Audio<br>• Number of Vehicle Antennas<br>• Patrol Speed if Moving Mode<br>• Make, Model, Year |

Table 6.1-2 -- **Additional Information to Collect**

# Chapter 6.2 -- Electronic Countermeasurers (ECM)

## Scams

BAD INFORMATION

One of the first attempts to fool traffic radar was to hang reflective objects on a vehicle, including dangling loose chains, putting foil or steel marbles in hubcaps, and securing strips of foil to the radio antenna.  None of these counters did anything to fool traffic radar and, in fact, may have enhanced the vehicle's radar cross-section enabling the radar to detect the vehicle at longer range! Even so drivers that used these techniques were subject to arrest in Rochester, New York, in 1953.

False information circulated on the Internet through massive emails in 2000 and 2001. Basically the scam message falsely claimed by over paying a ticket by a few dollars, mail-in only, and then not cashing a check for the difference when corrected would save you from losing the points.  The emails claimed the system would not charge the points until the check was cashed - *this was not* and *is not true* and *does not work* !

QUESTIONABLE PRODUCTS

Products come and go that claim to interfere with microwave police radar.  Be especially skeptical of **jammer** claims, and in particular any ***passive jammers***.  All known passive jammers are useless.  Some test indicate not only do these jammers not work, but may enhance the vehicle's radar cross-section.  A jammer antenna is a great reflector enabling radars to detect the target at longer range, 10 to 30 percent in some cases.  Even so drivers that use passive or active jammers are subject to FCC penalties.

Some hood covers, bras, and license plate covers are suppose to absorb and disperse **microwave** radar energy thus reducing vehicle detection range.  Any reduction is small with only a slight, if any, degradation in radar maximum detection range.

Some license plate covers supposedly disperse **laser** light denying a laser radar reflections from the license plate.  However if the laser is not aimed at the license plate or the beam is wide enough to overlap other parts of the vehicle this method will fail.  In fact most reports indicate license plate covers have little if any effect on laser, or microwave, radar performance.

# Radar Detectors

Some radar detectors will identify a radar signal sooner than others, but the difference is usually insignificant.  If a radar is continuously transmitting a good detector can give a mile or two warning under ideal conditions.  Bends in the road or objects that mask the signal will reduce detector performance as well as radar performance.  Radar detectors require at least 150 milliseconds, 0.15 seconds, to detect and identify a radar signal.

## INSTALLATION

One can get the most from any radar detector by proper installation.  The detector antenna should be clear of obstructions, especially metal such as windshield wiper arms, and mounted as **high as possible**.  Some electrically heated windshields have metal film coatings reflective to microwaves, these windshields obstruct signals and prevent the use of a radar detector.  These windshields also increase a vehicle's radar cross section increasing s radar's detection range.

## LEGALITY

The Federal Communication Commission (FCC) does not regulate radar detector legality, only detector emissions (FCC DA 96-2040).

Radar Detectors are illegal in Virginia, Washington D.C., all U.S. military installations, some provinces in Canada and some states in Australia.  The United States Federal Motor Carrier Safety Administration Regulation 392.71 prohibits use or possession of a radar detector in commercial motor vehicles and vehicles over 10,000 pounds.

**Radar Detector Prohibitions**
- Vehicles over 10,000 pounds
- Commercial Vehicles
- Military Installations
- Virginia
- Washington D.C.
- Some States in Australia
- Some Provinces in Canada

## RADAR DETECTOR OPTIONS

There is a large selection of traffic radar detectors with a variety of options and prices. Some features and options are listed below, not all detectors have all capabilities.

| Police Radar Transmissions | • X Band:    10.500 - 10.550 GHz<br>• K Band:    24.025 - 24.250 GHz<br>• Ka Band:  33.400 - 36.000 GHZ<br>• IR Band:   904 nm (laser) |
|---|---|
| Safety Message Transmissions | • Safety Warning Systems (SWS)<br>• Safety Alert Systems |
| Detector Detector Leakage | • VG-2 Interceptor<br>• Stalcar and Spectra series<br>• MD-3, RD-1, RD-2, RD-3 series |
| Strobe Light Detection | • Detects strobe light used by emergency vehicles to change stoplights. |

Table 6.2-1 -- **Electronic Signal Detection**

Most radar detectors sold in the USA detect X, K, Ka and IR laser radar signals. Some radar detectors can decode one or both of the Safety Message signals, SWS and Safety Alert. Radar Detector Detector (RDD) leakage and Strobe Light detection are optional features.

Some middle eastern countries and some European countries use Ku band radars that transmit at 13.50 GHz. Some European countries use X band radars at 9.41 GHz or 9.90 GHz.

| Coverage | • Forward<br>• Rear optional<br>• Sides (360°) optional |
|---|---|
| Radar Detected Indicators | • Audio Beeps - faster beeps stronger signal<br>• Radar Band Lights - X, K, Ka, or IR<br>• Signal Strength Bar or Meter optional<br>• Radar Frequency optional |
| Power Source<br>varies with model | • Vehicle DC plug<br>• Internal Battery |
| Countermeasures | • Shuts down on Detector Detector detection optional |
| Additional Features optional | • Compass - magnetic north<br>• Altimeter - altitude above sea level<br>• Weather Radio - NOAA VHF broadcast<br>• Temperature Outside Vehicle<br>• GPS<br>• Driver Alert - beeps periodically |

Table 6.2-2 -- **Standard Features and Options**

## Safety Messages / Warnings

There are 2 types of safety and warning message systems, Safety Warning Systems (SWS) and Safety Alert Systems.   Both systems transmit in the police radar K band. Radar detectors not designed to detect the safety message interpret the signal as a K band radar.   Some detectors detect a safety message warning signal, but not the message.  The SWS  has around 60 messages, the obsolete Safety Alert has 3 messages.

### Safety Warning System (SWS)

Radar detectors with SWS decoders have been available since 1996.  SWS receivers look for a safety warning signal message every 3 to 6 seconds.   Sixty-four short fixed text messages are possible, up to 2 different messages may be broadcast one after the other. Variable messages are allowed if none of the available fixed text messages is appropriate.

Messages can be broadcast from fixed positions to broadcast travel, weather, train approaching a crossing, construction information, or mobile vehicles including emergency vehicles, school buses, oversized vehicles.   Some stationary transmitters can be programmed from a remote location by radio or phone link.

### SWS Transmitters

Frequency:      24.100 GHz ± 50 MHz

Power (ERP):   50 milliwatts

Beamwidth      23° horizontal

### Safety Alert (obsolete)

Safety Alert System was introduced in the 1990's and transmits 1 of 3 fixed messages, Emergency Vehicle, Train, Road Hazard, using 24.07,  24.11, and 24.19 GHz.

## Jammers

### Microwave Jammers

From time to time traffic radar jammers appear in magazine and internet ads. Most if not all of these jammers are useless, absolutely no effect at any range under any conditions.

The legality of jammers is also in question and somewhat up to the whims of bureaucrats and politicians regarding enforcement and court interpretation of the law. Several U.S. states and some countries prohibit the use or possession of a police radar jammer.

For the United States Federal Communications Commission to consider an intentional radiator legal the field radiation must meet FCC Rules Part 15, *and* the device must perform some function for the public good. Police radar jammers are not considered good for the public by the FCC.

> The FCC considers the use of traffic radar jammers as malicious interference and strictly prohibited by the Communications Act of 1934, as amended, as well as by FCC rules. Anyone using a jammer risks such penalties as losing FCC licenses, paying a fine, or facing criminal prosecution.

FCC DA 96-2040.

### Passive jammers

Passive jammers are suppose to re-radiate the radar signal after distorting it, adding noise or rapidly shifting frequency, in such a way the true target reflection is masked. A passive jammer does not generate or amplify a signal, only channel or redirect the radar energy back toward the radar.

For this method to work the jammer antenna would need to be at least as large as the vehicle - not practical. Additionally the jammer antenna would need to be almost perfectly aligned to the radar antenna - not lightly. Passive jammers have absolutely no effect on any radar under any circumstances.

In 1997 the FCC ruled passive jammers violate federal regulations because the jammers radiate energy that, or at least is intended to, adversely affect the ability of law enforcement officials to protect public safety on the highways. Before this ruling passive jammers were not considered transmitters and not covered by FCC regulations.

### Active jammers

There are 2 types of active jammers, one continuously transmits, the second only transmits when a signal detected. Many police radars can detect jamming signals even when the radar is not transmitting, this is a good reason a jammer should only transmit when a radar signal is present.

| Continuously Transmitting | Transmits when Signal Detected |
|---|---|
| May set off Interference or Jamming Indicators. | Radar may get speed reading before jammer can react. |

There are 2 basic jamming techniques used against police radar, noise jamming and deceptive jamming. Noise jamming transmits amplified noise to desensitize the receiver and obscure legitimate targets. Deceptive jamming transmits a signal that causes the radar to register a speed set by the jammer. Both techniques could set off the radar interference or jamming indicators.

| Noise Jamming | Deceptive Jamming |
|---|---|
| Desensitizes radar receiver obscuring legitimate targets. | Transmits a deceptive fake speed fixed or set by operator. |

Deceptive jamming transmits a signal anywhere in the radar receive band. The signal is slightly amplitude modulated at the *radar Doppler frequency rate of the fake speed*. This technique works on the same principle as a tuning fork.

## Laser Jammers

Most laser jammers are completely ineffective. A few jammers are effective, about half the time, for vehicles at ranges greater than 500 to 1000 feet. All jammers are completely ineffective at ranges less than 500 feet. The jammers have the same problem as laser detectors, the narrow lidar beam never strikes the detector aperture. Laser Jammers are useless at the ranges lidar is most effective.

Laser jammers are ineffective because the jamming signal must closely match the lidar pulse width and rate which varies with model. The jammer must detect the waveform and generate a match, then synchronized transmission to the next detected pulse.

## Radar Detector Detectors (RDD)
Electronic Counter Countermeasures (ECCM)

A radar detector detector (RDD) is used to enforce anti-detector laws by sensing signals that leak from a radar detector's local oscillator (LO).  Detecting a signal indicates a radar detector is operating near-by.

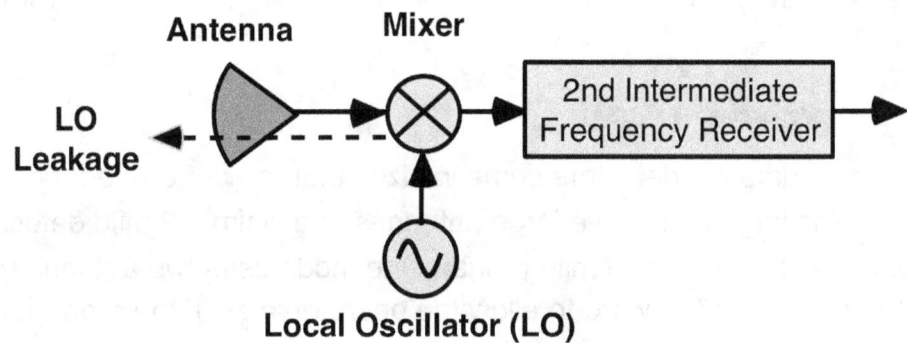

Figure 6.2-1 -- **Radar Detector Local Oscillator Leakage**

A radar detector's mixer combines the local oscillator signal with received signals to convert to an intermediate frequency for processing.  Mixers are not perfect and leak some of the LO signal to the antenna.  Some radar detectors have a pre-amp or isolator between the mixer and antenna that should reduce LO leakage, but may not eliminate it.  The old and unreliable crystal radar X band detectors cannot be detected by radar detector detectors because the they do not have a local oscillator.

Some mobile amateur radio transceivers leak signals in the X band that can set off an RDD.  Several people operating legal ham radios were wrongfully accused of possessing radar detectors because their radios were detected by an RDD.

RDD's also have LO leakage, and some radar detectors are designed to detect this leakage.

## Radar Detector Detector Models

### Stalcar and Spectre series
Stealth Micro Systems Pty Ltd., Brisbane, Queensland (Australia)

The first Stalcar series was introduced in 2000.  Years later the Stalcar was replaced by the Spectre series. The units are about the size of a radar detector with about half a mile range.   The receivers use double down conversion superheterodynes, similar to police radar receivers.  Additionally Varactor Tuned Gunn oscillators allow the receiver to sweep a large band of frequencies.   Frequency bands increased from 11 - 15 GHz to 10 - 25 GHz.  Receiver sensitivity is on the order of a police radar sensitivity, -110 dBm.

### RD series
Hill Country Research, Fredericksburg, TX (USA)

The RD series radar detector detectors come in sizes that range from radar detector size to suitcase size.  The large units have large antennas and claim a 2 mile detection range. The smaller units have about a half mile range.  One mode uses two antennas, one aimed forward and the other 45° off forward to allow the operator know if the radar detector is in front or off to the side.

### VG-2 Interceptor
Technisonic Industries Ltd., Missasauga, Ontario (Canada)

The VG-2 Interceptor radar detector detector was introduced in the 1980s.  It has a large antenna and is about the size of half a shoebox.  Detection range varies from 1/4 mile to 2 miles.  The VG-2 is tuned to receive microwave signals between 11.4 and 11.6 GHz.  At one time 11.558 GHz was a universal radar detector LO frequency.  Today radar detectors come in numerous LO frequencies, most are out of the VG-2's band.

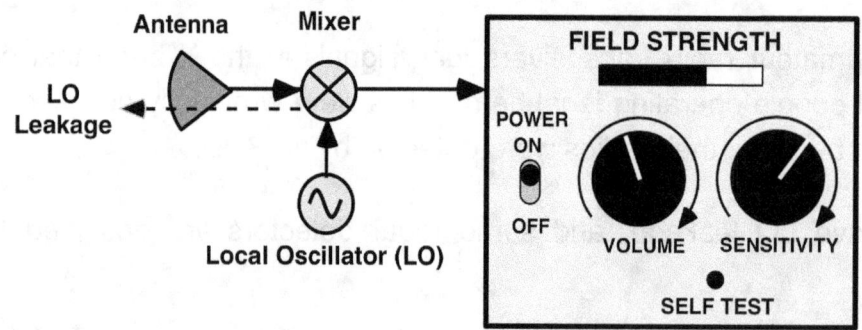

## Chapter 6.3 -- The Courtroom

CAVEAT: This document is not intended to give legal advice in any way. This section and any other sections or references to the law or courtroom activities is based solely on the author's personal experiences and observations.

All technical information and conclusions in the *Police Radar Handbook* is quantifiably described using illustrations, graphs, tables, or mathematical formulas -- based on or derived from fundamental scientific and engineering principles, published factory specifications, measured data, or U.S. Government documents.

## Radar and Speed Limit Requirements

### Radar FCC License

Microwave police radars are regulated by the Federal Communications Commission (FCC) Rules and Regulations, Part 15, 20, and 90, radiolocation services. The FCC specifics technical standards such as operating frequency, bandwidth, power density, etc. The rules do not cover the calibration of radar units, radar accuracy, or operator capability requirements.

| License Requirements |
| --- |
| State or local government agencies (including police) that have an FCC license for a communication system (Public Safety Radio Services) are not required to have a separate FCC license for traffic radar under part 90 of FCC rules. |
| Radar units may also be used under Part 90 (other appropriate FCC radio license required) by non-public safety entities such as professional baseball teams, tennis clubs, automobile and boat racing organizations, private transportation firms, railroads, etc., to measure the speed of objects or vehicles. |
| Public safety agencies can operate unattended, low-power, transmit-only radar units. |
| Non-public safety users are required to obtain a Part 90 license. |

Police do not need an FCC license to operate traffic radar if their radios are licensed, however other state, local, or agency requirements may apply.

## Laser Radar

Some states limit or restrict laser radar use. In New Jersey troopers can only use laser radar in clear weather and for vehicles less than 1000 feet.

## NHTSA / IACP Consumer Product List (CPL)

The National Highway Traffic Safety Administration (NHTSA), in conjunction with the International Association of Chiefs of Police (IACP) and the Law Enforcement Standards Laboratory of the National Institute of Standards and Technology (NIST) developed testing protocols and performance standards for speed measuring devices. Radars, microwave and laser, that meet the standards are included in the NHTSA / IACP Consumer Product List (CPL) of approved speed measuring devices.

Time / distance measurement features not approved or tested. Moving radars approved only for automobile or truck use, not motorcycles, boats, etc.

CPL approval insures that a microwave or laser radar model, at least the unit tested, meets basic minimum standards set by the NHTSA, IACP, and NIST. Some states and agencies require speed measuring devices be CPL approved.

## The Manual on Uniform Traffic Control Devices (MUTCD)

The Manual on Uniform Traffic Control Devices (MUTCD), established by federal law in 1966, sets national standards for highway and road issues. *The MUTCD requires speed limits are to be determined by an Engineering Study as defined by MUTCD 1A.13.* An Engineering Study must also be done before any speed limit can be changed. If some government body changes a speed limit without a proper study, the speed limit is illegal.

## Take it to Court?

Police know the uniform, flashing lights, siren, car markings, loaded guns, and the power to arrest are intimidating factors when issuing a speeding ticket. Some officers are intentionally intimidating to help insure tickets are not challenged. Don't let intimidation be a factor in determining whether to challenge a ticket. Do keep in mind the odds of successfully beating a radar ticket, innocent or not, are against the defendant, but not zero.

Several factors must be considered before deciding to fight a speeding charge. First and most important one must be wrongly accursed, do not try to beat a charge if you're guilty. However, if you were not speeding as fast as accused, you might plead for the lesser speed violation to lower the fine if you can convince the judge. With some insurance companies the higher the speed the more the insurance rate increases on conviction. The court clerk should know if the fine would change for a lower speeding ticket, and your insurance agent should know if rates change with speed.

In general the more people the court serves the better chance for a fair hearing. A small city or county court is usually the hardest to prove a case and in some situations it is virtually impossible to walk away without paying something.

Some courts require the defendant to plead guilty or not guilty in a pre-trial hearing, at which time a trial date is set for a not guilty plea. This means two court appearances. Some courts let defendants mail in the plea so only one court appearance is necessary.

## Preparation for Court

Be prepared to document as many facts as possible.  Ground and aerial photographs of the site could be helpful. Show approximate distances from the map scale or actual measurements.  Note approximate locations for vehicle and radar at start and finish of  the radar track.  Be careful any maps or photographs are not outdated.

## Discovery

Some experts suggest filing a Discovery motion or a request for Discovery form with the court clerk for some or all of the following items;

| | |
|---|---|
| **Video** | • Dash Camera or other Video / Audio |
| **Radar Documents** | • Make, Model<br>• Manufacturer Certificate of Calibration<br>• Operator Manual and Specifications<br>• Calibration Test Log Sheets |
| **Tuning Fork Documents**<br>not laser radars | • Resonance and Radar Frequency<br>• Tuning Fork Calibration Logs |
| **Officer Notes** | • Notes of incident and log book |
| **Officer Training** | • Certificate of Competency<br>• Training Materials<br>• Training Classes |
| **Department Policies** | • Radar Standards and Practices<br>• Test Procedures |

Table 6.3-1 -- **Discovery Motion Items**

You or your lawyer must go through the court clerk to request discovery information by filling out a form, or making a formal request in writing regarding specific information.  In most cases an additional processing and copy fee is required.   If the officer cannot produce a reasonable request for information the charge may be dismissed.

Some schools of thought suggest that requesting information might negate any chance of the officer not showing up for court in which case the charge should be dismissed.

## Officer Presentation

Most officers are trained to collect certain information in the event the case is disputed. Below list typical minimum information an officer *should* be prepared to present in court.

---

1. Establish time, place, location of radar.

2. Establish location of offending vehicle.

3. Identify the offending vehicle.

4. Establish and identify vehicle operator.

5. Visually observed apparent excessive speed.

6. Observed vehicle was alone out front, or fastest for Fastest Mode.

7. Established radar tested before use.
   • Self-Test run before and after use.
   • Tested against vehicle with calibrated speedometer.

   **Microwave Radar**          **Laser Radar**
   • Tested with Tuning Forks.   • Beam alignment tested.
                                 • Range Accuracy tested.

8. State qualifications and training

---

Table 6.3-2 -- **Officer Courtroom Presentation**

## Defendant Presentation

When disputing a speeding charge one should already know the answers to most of the questions listed below. If for any reason some or all of the pertinent information is not known in advance, ask the officer in court under oath.  Do not ask questions the officer has already addressed unless disputed. The judge will side with the officer on most disagreements such as exact locations, vehicle placements and distances unless one has hard proof.

## Documentation

| Radar / Lidar Documentation | Laboratory Calibration |
|---|---|
| • Is the radar factory certified?<br><br>• Is the radar CPL approved?<br><br>• Make, Model, Year purchased?<br><br>• Hand held or fixed mounted? | • Radar / Lidar in Calibration?<br>  - Last and next calibration test?<br>  - Who conducts calibration test?<br><br>• Microwave Radar<br>  - Tuning forks last and next calibration?<br>  - Who conducts tuning fork calibration? |

## Operator Testing

| Guidelines | Tests |
|---|---|
| • What are operator **Testing** Guidelines?<br>  - When and how often **self-test** run?<br>  - **Site pre-tested** with test vehicle?<br><br>• **Operation** Guidelines<br>  - **Minimum time** for steady stable track?<br>  - **Radar distance from traffic lanes?**<br><br>• Who established guidelines? | • Microwave Radar<br>  - Radar tested with tuning forks?<br>  - **Demonstrate test procedure**?<br><br>• Laser Radar<br>  - Range accuracy tested, what range?<br>  - Horizontal and Vertical Alignment tested?<br>    What range? |

## Radar Information

| Specifications | Modes Available |
|---|---|
| • Radar Band?  X, K, Ka, IR | • Stationary and Moving modes? |
| • Beamwidth? | • Fastest mode? |
| • Speed measuring range min to max? | • Opposite and same-lane traffic modes? |
| • Maximum detection range? | • **Range Control**? |
| • Sample time for 1 speed reading? | • **Patrol Speed Blanking**? (moving mode)<br>  - Purpose and use? |

## Officer Training and Events

| Officer Training | Events |
|---|---|
| • Hours of radar training? | • **Radar distance from traffic lane**? |
| • Who trained? | • Lanes covered by radar beam? |
| • Time in radar enforcement? | • Range vehicle first detected? |
| • Speeding citations issued that day? | • **Time** vehicle tracked? |
| • Number of citations disputed? | • Steady stable signal? |
| • Quota system in agency? | • Any Cosine Effect errors observed? |
| | • **Range Control adjusted for conditions**? |
| | • Patrol vehicle or other transmitters on? |
| | • Total transmitters on patrol vehicle? |

Not all questions or details are appropriate for all cases.  The better you are prepared, the better the odds of a favorable outcome.  The more pertinent questions you can ask that the officer cannot answer, the stronger your case.

## Court Day

On court day be prepared to spend the entire day. Do not be surprised that typically 100 or more people are scheduled at the same time. Before a trial one of the prosecutors may try to talk you into pleading guilty. Be prepared for scare tactics, intimidation and time-consuming distractions designed to make one believe pleading guilty is better than going to trial. No plea bargain, no deal.

If the prosecutors believes your case is strong you may be offered a plea bargain. The stronger you make your case the better the plea bargain, but do not tip too much information in the event you go to trial. Sometimes you can bargain the plea bargain, sometimes it's take it or leave it. Bargaining to pay the fine and to not have the ticket on record for the insurance company to see is sometimes the best deal, sometimes not. Most plea bargains also have a probation period of 30 to 90 days, if you don't get any tickets during probation the ticket goes off the record.

If you do go to trial forget the prosecutor, now the judge is the only one to persuade. Don't try to make a case with material you don't thoroughly understand, most judges recognize a shaky case. Do not be afraid to ask the officer specific, pertinent and detailed questions about the radar. The more questions the officer cannot answer, the less creditable the prosecution's case. An officer that does not have creditable working knowledge about radar should not be operating one.

The general format for traffic court is the prosecution will state their case and question the officer for support. You will then have a chance to state your case and question the officer. The prosecutor will then try to shatter your case by questioning you. Be aware judges have authority to conduct court procedures to their satisfaction, this means court procedures vary with state, county, city, judge, magistrate and arbitrator.

# The Cosine Effect Defense

## Minimum Range

The Cosine Effect works in favor of the motorists in terms of measured speed, a little lower than actual speed.  The Cosine Effect also puts some limitations on the radar, mainly a minimum range.  Vehicles less than minimum range cannot be measured.  The greater the distance the radar from the traffic lane, the greater the minimum range.  A vehicle inside minimum range could be mistaken for a vehicle just outside minimum range.

## Large Cosine Effect Angle

A common occurrence is the radar cosine effect angle is so severe when measured speed is corrected for the angle the corrected speed is excessively high and unlikely.  This would indicate the radar was tracking another signal such as another vehicle or interference from a transmitter in or near the patrol vehicle.

For example if the cosine angle is 60° radar measured speed is half true speed.  If the officer states observed estimated vehicle speed to be about the same as the radar measured speed, one could argue the officer's judgment was biased incorrectly by the radar.  If the officer's estimated speed conflicts with radar corrected speed, which one is correct, if any?

## Observed Cosine Effect

The cosine effect causes measured speed to change, the closer the target vehicle the lower the measured speed.  A good steady stable track will show cosine effects as the vehicle gets close to the radar, measured speed decreases and stops at minimum range.

A 75 mph vehicle approaching a microwave or laser radar will display speed readings that look something like "75 ------------ 74 -------------------74 --- 73 -- 72".  The sequence is reversed for receding vehicles, speed increases.  Noticeable speed changes should only take place relatively close to the radar.  The speed readings are predictable and should be noticeable to the officer.

# Chapter 7 -- Frequency and Wavelength

## Chapter 7.1 -- Frequency Bands

| | | | | |
|---|---|---|---|---|
| hertz | Hz | cycle per second | $10^0$ Hz | 1 Hz |
| kilohertz | kHz | thousand hertz | $10^3$ Hz | 1,000 Hz |
| megahertz | MHz | million hertz | $10^6$ Hz | 1,000,000 Hz |
| gigahertz | GHz | billion hertz | $10^9$ Hz | 1,000,000,000 Hz |
| terahertz | THz | trillion hertz | $10^{12}$ Hz | 1,000,000,000,000 Hz |

Table 7.1-1 -- **Frequency Multipliers**

### Police Radar Frequencies

| Band | Frequency | Wavelength | | Use |
|---|---|---|---|---|
| S | 2.445 GHz | 4.827 in | 12.261 cm | obsolete |
| X | 9.410 GHz | 1.254 in | 3.186 cm | Europe |
| X | 9.500 GHz | 1.192 in | 3.028 cm | Europe |
| **X** | **10.525 GHz** | 1.121 in | 2.848 cm | **USA** |
| Ku | 13.450 GHz | 0.878 in | 2.229 cm | Europe, Middle East |
| **K** | **24.125 GHz** | 0.4892 in | 1.243 cm | **USA**, Australia, Europe |
| **K** | **24.150 GHz** | 0.4897 in | 1.241 cm | **USA** |
| **Ka** | **33.40 - 36.0 GHz** | 0.353 - 0.328 in | 8.976 - 8.328 mm | **USA**, Australia, Europe |
| **IR** | **331.6 THz** | 904 nanometers | | Laser Radar |

in - inches, cm - centimeters, mm - millimeters

Table 7.1-2 -- **Worldwide Police Radar Frequencies**

# Frequency Band Designations

## Military Radar Bands

Military radar band nomenclature, L, S, C, X, Ku, and K bands originated during World War II as a secret code so scientists and engineers could talk about frequencies without divulging them.   After the war the codes were declassified and Ka band and millimeter (mm) were added.  Military radar band nomenclature is widely used today in radar, satellite and terrestrial communications, and military  electronic countermeasure applications.

| Band | Frequency | Wavelength | | |
|------|-----------|------------|-----------|---|
| HF   | 3 - 30  MHz    | 100 - 10 meters | 327 - 32 feet    | High Frequency |
| VHF  | 30 - 300  MHz  | 10 - 1 meters   | 32.8 - 3.3 feet  | Very High Frequency |
| UHF  | 300 - 1000  MHz | 100 - 30 cm    | 3.3 - 1.0 feet   | Ultra High Frequency |
| L    | 1 - 2  GHz     | 30 - 15 cm      | 11.8 - 5.9 inch  | microwaves |
| S    | 2 - 4  GHz     | 15 - 7.5 cm     | 5.90 - 2.95 inch | |
| C    | 4 - 8  GHz     | 7.5 - 3.7 cm    | 2.95 - 1.48 inch | |
| X    | 8 - 12  GHz    | 3.7 - 2.5 cm    | 1.48 - 0.98 inch | |
| Ku   | 12 - 18  GHz   | 2.5 - 1.7 cm    | 0.98 - 0.65 inch | |
| K    | 18 - 27  GHz   | 1.7 - 1.1 cm    | 0.66 - 0.44 inch | |
| Ka   | 27 - 40  GHz   | 11.1 - 7.5 mm   | 0.44 - 0.30 inch | millimeter wavelengths |
| mm   | 40 - 300  GHz  | 7.5 - 1.0 mm    | 0.30 - 0.04 inch | |

Table 7.1-3 -- **Military Radar Bands**

Military HF, VHF, UHF same as Radio Band HF, VHF, UHF respectively.

## ITU Radar Bands

The International Telecommunications Union (ITU) specifies bands designated for radar systems. The ITU radar bands are sub-bands of military designations.

| ITU Band | Frequency Band | |
|---|---|---|
| VHF | 138 - 144<br>216 - 225 | MHz |
| UHF | 420 - 450<br>890 - 942 | MHz |
| L | 1.215 - 1.400 | GHz |
| S | 2.3 - 2.5<br>2.7 - 3.7 | GHz |
| C | 5.250 - 5.925 | GHz |
| X | 8.500 - 10.680 | GHz |
| Ku | 13.4 - 14.0<br>15.7 - 17.7 | GHz |
| Ku | 24.05 - 24.25 | GHz |
| Ka | 33.4 - 36.0 | GHz |

Table 7.1-4 -- **International Telecommunications Union Radar Bands**

VHF - Very High Frequency
UHF - Ultra High Frequency

## Radio Bands

| Band | Nomenclature | Frequency | Wavelength |
|------|-------------|-----------|------------|
| ELF | Extremely Low Frequency | 3 - 30   Hz | 100,000 - 10,000   km |
| SLF | Super Low Frequency | 30 - 300   Hz | 10,000 - 1,000   km |
| ULF | Ultra Low Frequency | 300 - 3000   Hz | 1,000 - 100   km |
| VLF | Very Low Frequency | 3 - 30   kHz | 100 - 10   km |
| LF | Low Frequency | 30 - 300   kHz | 10 - 1   km |
| MF | Medium Frequency | 300 - 3000   kHz | 1,000 - 100   m |
| HF | High Frequency | 3 - 30   MHz | 100 - 10   m |
| VHF | Very High Frequency | 30 - 300   MHz | 10 - 1   m |
| UHF | Ultra High Frequency | 300 - 3000   MHz | 100 - 10   cm |
| SHF | Super High Frequency | 3 - 30   GHz | 10 - 1   cm |
| EHF | Extremely High Frequency | 30 - 300   GHz | 10 - 1   mm |

Table 7.1-5 -- **Radio Frequency Bands**

km - kilometers
m - meters
cm - centimeters
mm - millimeters

## ECM Bands

The electronic countermeasures (ECM) industry has it's own band designations.

| Band | Frequency Band |
|:---:|:---:|
| A | 30 - 250  MHz |
| B | 250 - 500  MHz |
| C | 500 - 1000  MHz |
| D | 1 - 2  GHz |
| E | 2 - 3  GHz |
| F | 3 - 4  GHz |
| G | 4 - 6  GHz |
| H | 6 - 8  GHz |
| I | 8 - 10  GHz |
| J | 10 - 20  GHz |
| K | 20 - 40  GHz |
| L | 40 - 60  GHz |
| M | 60 - 100  GHz |

Table 7.1-6 -- **ECM Bands**

## Sound Waves

Sound waves are air pressure waves that travel at 765 mph at sea level, not like electromagnetic radio or radar waves that travel at the speed of light.  Sound is a pressure wave of vibrating air molecules, and does not exits in the vacuum of outer space.

Most people at best can hear sound waves between 20 and 20,000 Hertz, the audio band. Sound, pressure waves, can extend as high as 10 MHz, however above 160 kHz propagation range greatly decreases due to absorption by atmospheric gases, air.

| Band | Frequency |
|---|---|
| infrasound | 0 - 20 Hz |
| audio | 20 - 20,000 Hz |
| ultrasound | 20 kHz - 10 MHz |

| Frequency | Use | Band |
|---|---|---|
| 0 - 20 Hz | Elephants, Whales | infrasound |
| **20 - 20,000 Hz** | **Human, Animals, Fish, SONAR** | **audio** |
| 10 - 30 kHz | Rodents | audio - ultrasound |
| 20 - 75 kHz | Insects | ultrasound |
| 20 - 160 kHz | Bats, Dolphins | audio - ultrasound |
| 0.1 - 2 MHz | Structures Test | ultrasound |
| 1 - 10 MHz | Medical Applications | ultrasound |

AM radio broadcast electromagnetic waves from 0.5 - 1.6 MHz

# Chapter 7.2 -- Electromagnetic Waves

## Electromagnetic Wave Parameters

Parameters that describe electromagnetic waves include frequency, wavelength and period. Frequency is cycles per second (Hertz), wavelength is distance traveled to complete 1 cycle and period is time to complete 1 cycle. The higher the frequency, the shorter the wavelength.

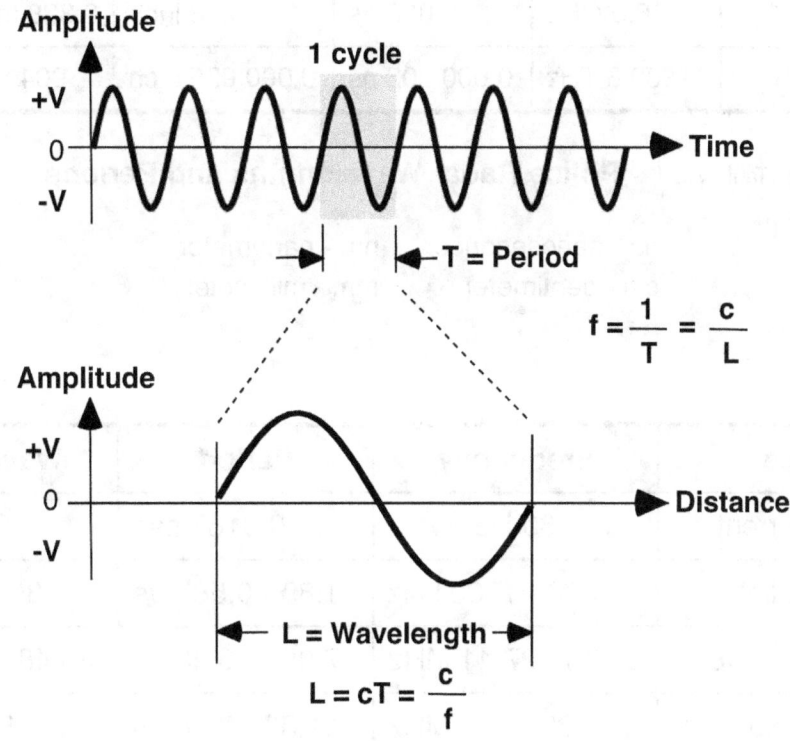

Figure 7.2-1 -- **Sine Wave Parameters**

$$f = c / L = 1 / T \qquad\qquad T = 1 / f \qquad\qquad L = c / f = c\,T$$

f = frequency        T = period, time to complete 1 cycle
L = Wavelength       c = speed of light

| Band | Frequency | Period | Wavelength | |
|------|-----------|--------|------------|---|
| S | 2.445 GHz | 0.409 ns | 4.83 inch | 12.261 cm |
| X | 10.525 GHz | 0.095 ns | 1.12 inch | 2.848 cm |
| K | 24.150 GHz | 0.041 ns | 0.49 inch | 1.241 cm |
| Ka | 33.4 - 36.0 GHz | 0.030 - 0.028 ns | 0.35 - 0.33 inch | 8.976 - 8.328 mm |
| IR | 333.6 THz | 0.000 003 ns | 0.000 035 inch | 904 nm |

Table 7.2-1 -- **Police Radar Wavelengths and Periods**

ns - nanoseconds     nm - nanometer
cm - centimeter      mm - millimeter

| Source | Frequency | Period | Wavelength |
|--------|-----------|--------|------------|
| House Current | 60 Hz | 0.0167 sec | 3,105 miles |
| AM Radio | 530 - 1700 kHz | 1.89 - 0.589 $\mu$s | 1,856 - 579 feet |
| Citizens Band (CB's) | 26.96 - 27.41 MHz | 37.09 - 36.48 ns | 36.48 - 35.88 feet |
| FM Radio | 88 - 108 MHz | 11.36 - 9.26 ns | 11.18 - 9.11 feet |
| DTV | 470 - 806 MHz | 2.13 - 1.24 ns | 25.11 - 14.64 inch |

Table 7.2-2 -- **Wavelengths for Common Electromagnetic Sources**

## Electromagnetic Wave Propagation and Polarization

## PROPAGATION

Electromagnetic waves consist of an **electric field** and a **magnetic field** *at right angles to each other*. The electric field, E-field, is measured in volts/meter. The magnetic field , H - field, is measured in amperes/meter. The electric and magnetic fields are analogous to voltage and current in circuits. The ratio of electric field to magnetic field in free space is **377 ohms**. Power is the vector cross product of the electric and the magnetic fields.

## POLARIZATION

Polarization is the plane of the electric field. In the USA over-the-air digital television signals are horizontally polarized while commercial AM and FM broadcast are vertically polarized. Radar waves may be polarized horizontally, vertically, angled, or circular.

Circular polarization is a technique where the electric and magnetic fields rotate at a rate proportional to frequency. Right hand polarization has the fields rotating in a clockwise direction looking from the antenna to the direction of travel, counter clockwise to a target. Left hand circular polarization is just the opposite.

Circular polarization tends to propagate better in rain, but has some signal loss in the polarization process. Circular polarized signals tends to be more reflective for some targets.

## Power Density, ERP, Electric and Magnetic Fields

Effective Radiated Power, ERP, is the most common unit for measuring signal propagation power. ERP is a function of transmitter power, signal loss to antenna, and antenna gain. ERP equals the product of antenna gain (G) and power delivered to the antenna ($P_t/L$).

### Effective Radiated Power (ERP)

$$P_{ERP} = P_t \, G \, / \, L$$

$P_{ERP}$ -- Effective Radiated Power (watts)      G -- antenna gain (ratio)
$P_t$ -- transmitter power (watts)                 L -- transmitter to antenna losses (ratio)

Power density, and electric and magnetic field strengths are a function of range.

| | | |
|---|---|---|
| Power Density ($P_d$): | $P_{ERP} / (4 \pi R^2)$ | W/m$^2$ |
| Electric Field (E): | $(P_d Z)^{0.5}$ | V/m |
| Magnetic Field (H): | $E / Z$ | A/m |
| Free Space Impedance (Z): | $E / H$ | ohms |

$P_{ERP}$ = Effective Radiated Power (watts)      $P_d$ = Power Density (watts / square meter)
E = Electric Field (volts / meter)                H = Magnetic Field (amperes / meter)
R = Range (meters)                                $\pi$ = Pi = 3.14159...
Z = Free Space Impedance = **376.73 ohms**

The electromagnetic field is a function of the inverse square ($1/R^2$) of range. Radar signals take a 2-way path, the loss is much greater and a function of the inverse to the fourth power ($1/R^4$) of range.

## MEASUREMENTS

Power density and electric and magnetic field strengths measurements must be measured in the antenna Far-Field.  The far-field is the range greater than twice the maximum antenna diameter squared divided by wavelength.

$$R_{min} = 2 D^2 / L$$

$R_{min}$ = Far-Field range,   D = Maximum Antenna Diameter,   L = Wavelength

$$R_{min} = 0.17 D^2 f_o$$

$R_{min}$ = Far-Field Range in inches,   D = antenna diameter in inches,   $f_o$ = Frequency in GHz

| Antenna Diameter | Antenna Far Field Range | | |
|---|---|---|---|
| | X Band | K Band | Ka Band |
| 1 inch | 2 inch | 4 inch | 6 inch |
| 2 inch | 7 inch | 17 inch | 25 inch |
| 3 inch | 2 feet | 3 feet | 5 feet |
| 4 inch | 3 feet | 6 feet | 8 feet |
| 5 inch | 4 feet | 9 feet | 13 feet |
| 6 inch | 6 feet | 13 feet | 19 feet |

Table 7.2-3 -- **Police Radar Far-Field**

Measurements taken in the near-field have absolutely no relation and do not apply to the far-field.  Radar manufacturers have a bad habit of listing power densities measured in the near-field, a meaningless number.   Just as bad power density measurements are sometimes listed without stating the range of the measurement, another meaningless number.

NOTES

# Chapter 8 -- Radio Frequency (RF) Health Issues

## Chapter 8.1 -- RF Biological Effects

### Radiation

Natural radiation was the only source of human exposure until the latter part of the nineteenth century when Thomas Edison invented the electric light. Most natural radiation of significance occurs in a small part of the lowermost frequency spectrum, electrostatic to about 5 kHz, and in the uppermost part of the spectrum above 10 THz or $10^{12}$ Hz. Man-made radiation dominates 50 Hz to 300 GHz.

Natural radiation below 5 kHz results from lightning. Average rate of global lightning strikes is about 100 bolts per second. Some natural radiation below 5 kHz results from pulsations in the earth's magnetosphere during intense solar storms causing the Aurora over one or both poles to light up.

Radiation is categorized as either non-ionizing or ionizing. Ionizing radiation has enough energy, high enough in frequency, to break atomic bonds by removing one or more electrons and creating a charged atomic particle. The higher the frequency the shorter the wavelength and the greater the energy and ionization.

Non-ionizing electromagnetic radiation is divided into three categories; electrostatic or non time-varying, low frequency such as house current, and RF or radio frequency. Natural forms of electromagnetic radiation occur above 10 THz and includes infrared heat, visible light, and ionizing radiation such as ultraviolet, X-rays, and gamma rays. All forms of radiation can have adverse health effects when intense enough or time exposure long enough.

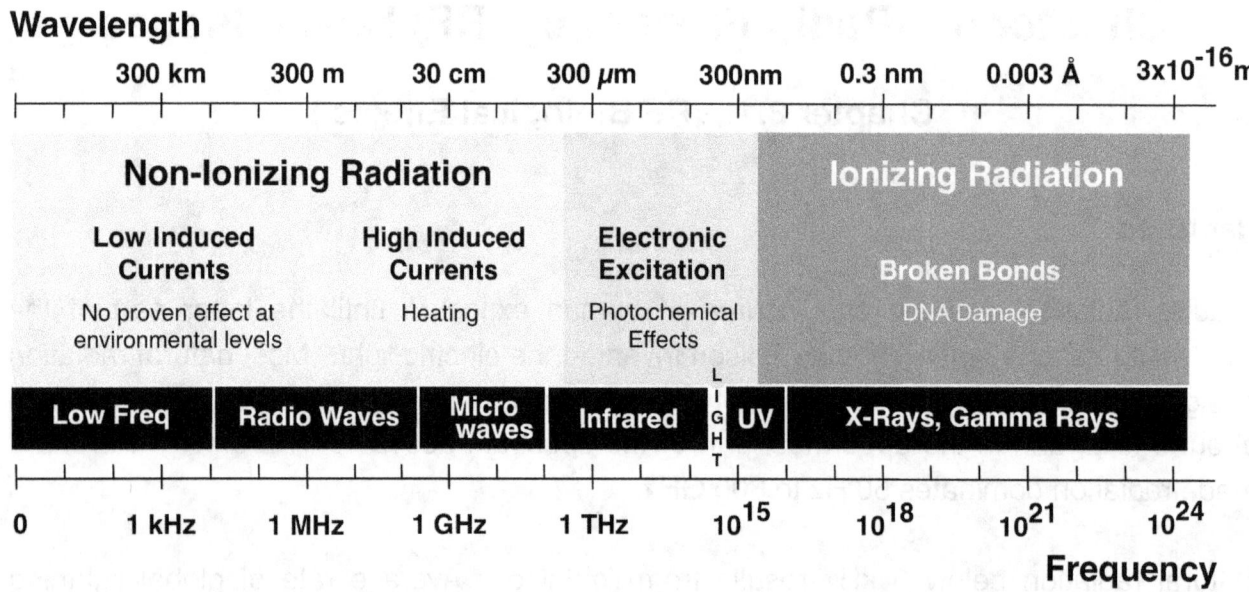

Figure 8.1-1 -- **Non-Ionizing and Ionizing Radiation**

## Studies and Reports

Police microwave radar has been linked to several adverse health effects by a number of police officers who operated radars over a long period, hours at a time for years. To date the scientific community cannot establish a mechanism that links police radar to adverse health effects, that does not mean one does not exist.

Many studies on health effects and exposure to electromagnetic fields conducted since 1948 have reached varying and sometimes contradictory conclusions. Much of the early research concentrated on the thermal heating affects of electromagnetic radiation, some later studies indicated reactions to electromagnetic fields not explained by thermal heating. Experts do not always agree on the levels or types of electromagnetic fields that affect health.

## US AIR FORCE STUDY

The U.S. Air Force sponsored a study of rats exposed to pulsed microwaves at 24.5 GHz. Police K band radar operates near the study frequency at 24.15 GHz. The study used pulsed microwaves and police radar is continuous.

The research was conducted by the University of Washington School of Medicine in Seattle and published in 1984. The study showed a significant increase in malignant tumors and noted affects in the adrenal glands and the entire endocrine system. The results suggests the maximum allowable exposure for humans is too low.

## 1998 STUDY

The London Times reported in 1998 that Dr. Henry Lai, an expert in non-ionizing radiation and professor at the School of Medicine and College of Engineering at the University of Washington, Seattle, announced that low-level microwave radiation can split DNA molecules in the brains of laboratory mice. DNA is Deoxyribonucleic acid, a complex, usually helical shaped chemical compound that is the substance that makes the organic matter of genes and chromosomes. Splitting DNA molecules in the brain is associated with Alzheimer's and Parkinson's Disease, and cancer. The cellular telephone industry supported Dr. Lai's research grant, but suppressed the report's publication.

## HEAT RAY

The Air Force Research Laboratory (AFRL) has developed a non lethal antipersonnel millimeter band *heat ray* intended for use on battle fields or against hostile crowds. A 3 x 3 meter, about 10 x 10 feet, antenna mounted on a Humvee, aircraft, helicopter, or ship can be swept across a crowd that induces skin heating. The system radiates a 2 second burst at 95 GHz, 3.16 millimeter wavelength, that can heat the skin to 130° Fahrenheit. Officials claim the energy only penetrates the top 1/64 inch (0.4 mm) of the skin and is not harmful to internal organs, no mention about the eyes. Operating range is around 700 to 1100 yards, rain, fog and humidity will reduce range. Possible countermeasures include shielding the energy using very thick clothing, a metallic sheet such as aluminum foil, or a metal trash can lid.

## Electrical Properties of Living Matter

Living matter exhibits many electric properties as well as generates various, relatively small, electromagnetic fields.  Medical doctors use known and well-documented electrical properties of the body to determine health and diagnose problems.  This section is a brief sample of some electrical properties of living matter.

• Nerve fibers consist of cylindrical membranes with one conducting fluid inside another fluid with a 0.1 volt potential difference between fluids.  A pulse causes the membrane between the fluids to temporarily become more permeable to ions and the voltage drops.  A pulse travels approximately 98 feet per second or about 67 miles per hour.

• Mechanical energy from bone bending and stress creates weak electrical potentials of a few millivolts across a centimeter at relatively low frequency.

• Electrocardiographs (EKGs) measure voltages between the chest and back to study heart functions.  The human heart also has an electric field near the surface between 1 and 10 volts per meter.

• Electroencephalographs (EEGs) measure potential differences on the order of microvolts in the scalp and are a meager measures of brain functions.  EEG patterns, brain waves, are different for every person, similar in twins, and similar for certain brain disorders such as epilepsy, brain tumors, brain damage, encephalitis, systemic diseases (toxemia and diseases of liver and kidneys).  When resting but not asleep, the back part of the head will register alpha waves or rhythms at a frequency of 8 to 12 hertz.  Beta waves, 18 to 25 hertz, relate to sensory functions and are smaller in magnitude than alpha waves.  People in comas have patterns with 1 to 3 hertz rates near the damaged area of the brain.  Theta waves are 4 to 7 hertz, normal in infants and young children but abnormal in adults.

## Thermal Effects of RF Radiation

Electromagnetic energy is well known to cause thermal heating in living tissue. Microwave ovens use electromagnetic energy to heat and cook food. Microwave ovens, introduced by Raytheon in 1947, are basically magnetron oscillators, a radar transmitter cavity tube, operating at 2.45 GHz. Some microwave ovens introduced in the late 1990's operate at 5.8 GHz.

The amount of heating that takes place is a function of transmit power and duty cycle time. Maximum surface heating due to RF exposure of a typical human occurs at frequencies between 30 and 120 MHz.

Tissue heating depends on the frequency of the source and the dielectric constant, water content and thickness of the tissue. The more conductive the tissue the more energy absorbed and heat generated. It requires a relatively large amount of radiation to heat tissue. Radiation levels too low to produce heat may have other effects at a cellular level, although not all experts accept this. Some experts believe most non thermal health effects require much higher field levels compared to thermal heating effects.

Fields strong enough to cause heating require hundreds to thousands of watts. Localized heat of about 1 watt per kilogram (0.45 W/lb) can damage tumors. The temperature of the tumor is raised to between 43 and 45 degrees Celsius (109 and 113 degrees Fahrenheit). Fields that cause mild heating can promote tissue healing or relax muscles.

**Specific Absorption Rate** (SAR) describes the energy absorbed by tissues and is measured in power per mass, typically watts per kilogram (W/kg). Absorption is a function of tissue permittivity and conductivity, and frequency of radiator. Safe limits for human exposure are sometimes based on *whole body exposure averaged over 0.1 hours (6 minutes)*. Some agencies consider a rate of **0.4 watts per kilogram (0.18 W/lb)** a safe limit.

## Interaction of Fields and Biological Systems

Weak RF fields insufficient to cause heating but strong enough to induce peak potentials of 1 to 1.5 millivolts per centimeter (0.10 to 0.15 V/m) can promote healing of broken bones. Experiments have shown osteoporosis, the loss of bone mass, can be halted or reversed by pulsed RF radiation.  The shape and timing of pulses is extremely significant, and different, to promote bone healing or affect osteoporosis.

Some studies indicate there are reactions to RF exposure not explained by thermal heating.  Strong RF fields heat tissue by vibrating molecules of the tissue.  Weaker fields can induce electrical currents in or on tissue, the stronger the field the larger the induced current.  Electric and magnetic fields can produce weak Lorentz forces that may affect charged particles, ions, on a molecular scale.  The Lorentz force is the force on a charged particle in motion due to the presence of an electric and magnetic field, and may boost or inhabit cell chemistries by pumping ions.

Listed below is a sample of some observed affects of electromagnetic radiation on living cells.  Most if not all effects depend greatly on frequency, modulation, and magnitude of the field.

- split DNA molecules in the brains of laboratory mice
- Peral chain -- randomly suspended particles (such as fat globules and E. coli bacteria) align with each other in direction of field
- Non spherical particles (such as E. coli) line up either perpendicular  or parallel to electric field depending on frequency
- Particle movement.
- Change in natural shape of cells.
- Cell death from membrane damage.
- Fusion of cells.

RF radiation has been linked to biochemical effects, immunological effects, Alzheimer's and Parkinson's Disease, cancer, cataracts, EEG effects and behavioral changes, to name a few. Studies from the old Soviet military suggest that some frequencies and modulations cause behavioral changes in humans. In 1990 the U.S. military was reported to be planning microwave radiation experiments on animals to study behavioral effects.

Some health effects seem to take place in small bands or windows of frequencies, modulations, and magnitudes. For example, nerve tissues affected by continuous 60 hertz fields are unaffected by 55 or 65 hertz fields. Some experiments have shown for some narrow frequency bands and specific modulation types a smaller field affects cells more than a stronger field. Several factors should be considered to determine electromagnetic radiation exposure and include;

> Distance:         Strength greatly decreases with distance
>
> Exposure Time:    Exposed to large fields for short periods,
>                   or small fields for long periods.
>
> Field Type:       Power, frequency, modulation

Also see National Technical Information Service (NTIS) publication number PB95-261350, *Occupational Exposure of Police Officers to Microwave Radiation from Traffic Radar Devices.*

# Chapter 8.2 -- RF Radiation Standards

## American National Standards Institute (ANSI)

American National Standards Institute (ANSI) standard C95.1-1982 sets electric and magnetic field strength limits for the general public for frequencies between 300 kHz and 100 GHz.   Below 300 MHz the electric and magnetic fields must be accounted for separately.  The standard is based on whole body exposure averaged over 6 minutes.

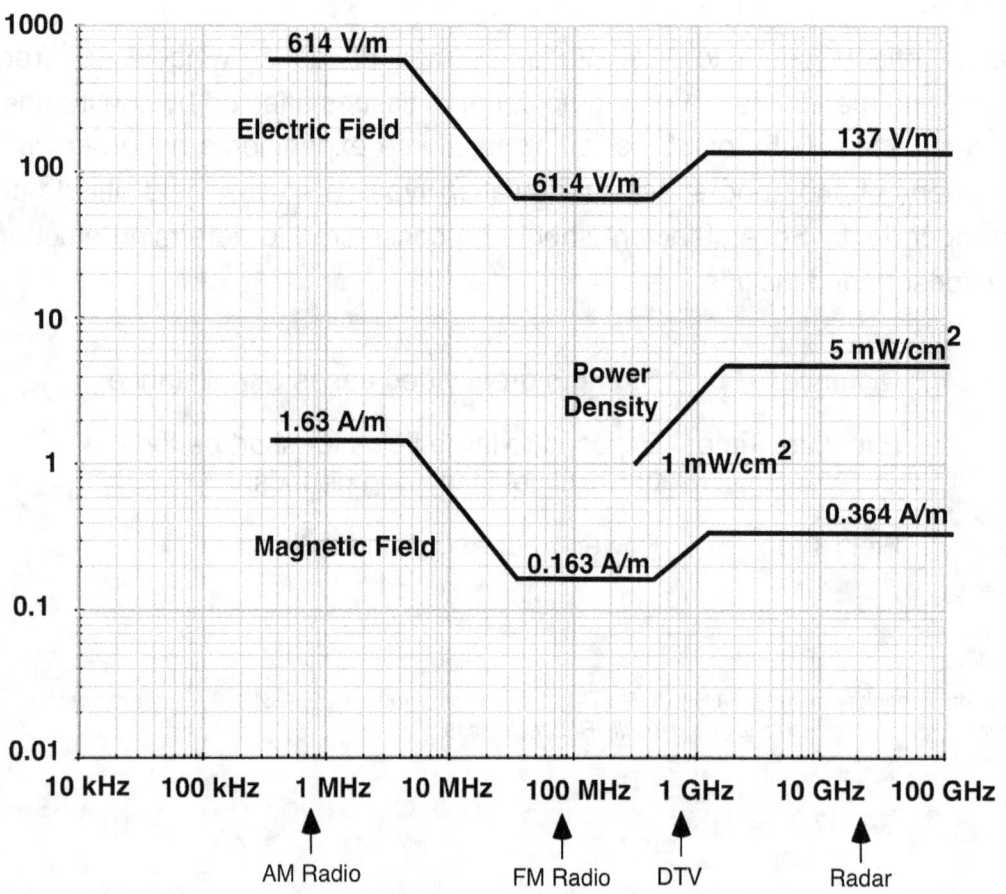

Figure 8.2-1 -- **ANSI Standard C95.1-1982**

At police radar frequencies (X, K, and Ka) the ANSI limit is 5 milliwatts/square centimeter. This equates to an electric field strength of 137 volts/meter and a magnetic field strength of 0.364 amperes/meter.

## Institute of Electrical and Electronic Engineers (IEEE)

The Institute of Electrical and Electronic Engineers (IEEE) standard C95.1-1991 sets electric and magnetic field strength limits for the general public for frequencies between 3 kHz and 300 GHz. Below 100 MHz the electric and magnetic fields must be accounted for separately. The standard is based on whole body exposure averaged over 6 minutes.

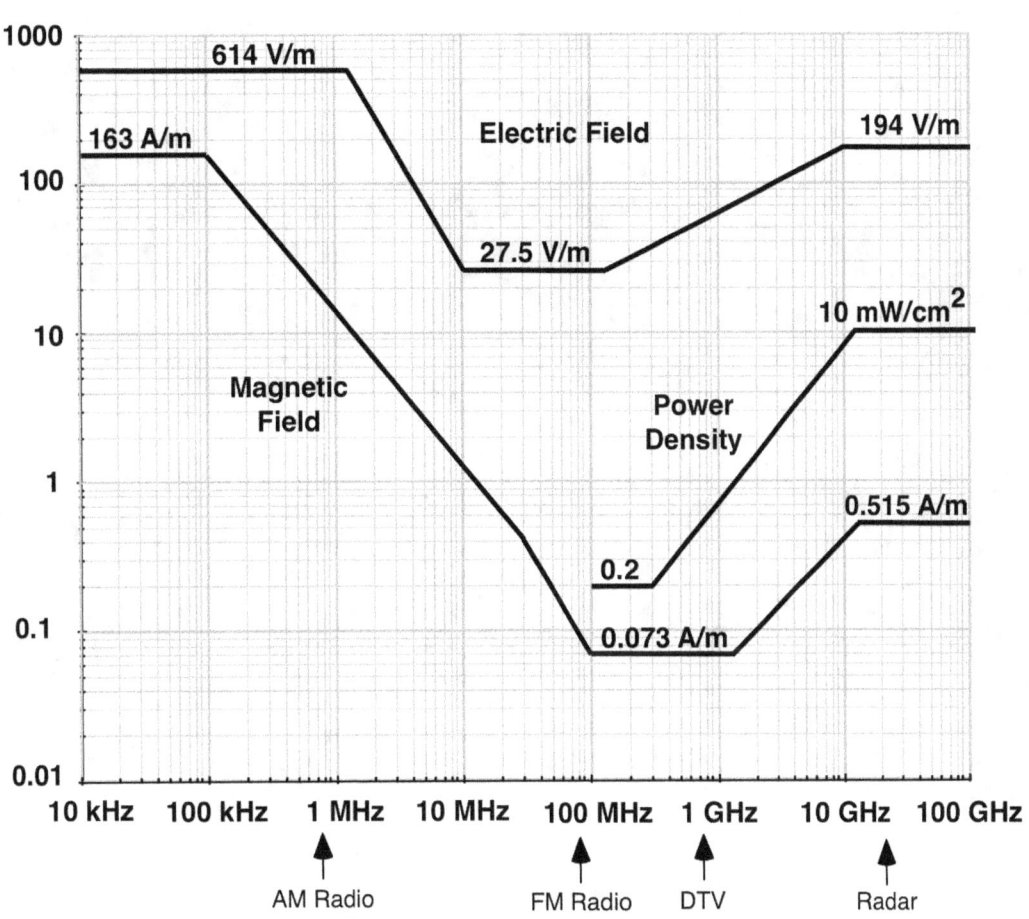

Figure 8.2-2 -- **IEEE Standard C95.1-1991**

At traffic radar frequencies (X, K, and Ka) the IEEE limit is 10 milliwatts/square centimeter. This equates to an electric field strength of 194 volts/meter and a magnetic field strength of 0.515 amperes/meter.

NOTES

# Appendix A -- Acceleration Parameters

## Acceleration due to Gravity

Objects falling due to the force of gravity in a free fall increase in speed with time and distance.  The exact acceleration rate varies slightly depending exactly where on earth the object is falling.  The earth is not a perfect sphere and gravity varies slightly depending on location.  By International definition 9.80665 meters per second second is the gravitational acceleration constant, $1g = 9.80665$ m/s$^2$ exactly.

| | | |
|---|---|---|
| **9.80665** | **m/s$^2$** | **meters per second per second** |
| 32.17405 | ft/s$^2$ | feet per second per second |
| 21.93685 | mph/s | mile per hour per second |
| 35.30394 | kph/s | kilometers per hour per second |
| 19.06260 | knots/s | nautical mile per hour per second |

Table A-1 -- **Gravitational Constant (g)**

Acceleration compared to free fall gravity acceleration g's is not a force. Force is mass multiplied by acceleration.  The term "g" is a fixed acceleration baseline for comparison purposes.

## Vehicle Acceleration and Braking

*Maximum braking* depends on vehicle weight and tire traction, width and diameter. Maximum *vehicle acceleration* depends on tires and horsepower. Top of the line production muscle cars can go from 0 to 60 mph in 5 seconds. Average acceleration is 60 mph per 5 seconds which equals 12 mph per second and equates to +0.55 g's.

| Time | 0 | 1 | 2 | 3 | 4 | 5 | seconds |
|---|---|---|---|---|---|---|---|
| **Speed** | 0 | 12 | 24 | 36 | 48 | 60 | mph |
| **Distance** | 0 | 9 | 35 | 79 | 141 | 220 | feet |

Table A-2 -- **0 to 60 mph in 5 seconds (+0.55 g's)**

Adaptive Cruise Control systems, vehicle radar, automatically brake when approaching other vehicles. Maximum braking varies from -3 to -5 meters/sec$^2$ (-7 to -11 mph/sec), or about -0.3 to -0.5 g's. Hard braking occurs around -0.55 g's.

| | | | |
|---|---|---|---|
| 0.1 g's | 3.2 ft/s$^2$ | 2.2 mph/sec | |
| 0.2 g's | 6.4 ft/s$^2$ | 4.4 mph/sec | |
| 0.3 g's | 9.7 ft/s$^2$ | 6.6 mph/sec | Moderate Braking |
| 0.4 g's | 12.9 ft/s$^2$ | 8.8 mph/sec | |
| 0.5 g's | 16.1 ft/s$^2$ | 11.0 mph/sec | Hard Braking |
| 0.6 g's | 19.3 ft/s$^2$ | 13.2 mph/sec | Skilled Driver |
| 0.7 g's | 22.5 ft/s$^2$ | 15.4 mph/sec | |
| 0.8 g's | 25.7 ft/s$^2$ | 17.5 mph/sec | Vehicle Maximum |
| 0.9 g's | 29.0 ft/s$^2$ | 19.7 mph/sec | |
| 1.0 g's | 32.2 ft/s$^2$ | 21.9 mph/sec | |

Table A-3 -- **Deceleration g's**

Many safety experts use 15 ft/sec² (0.47 g's) as the *maximum* deceleration that is safe for the average driver to maintain control, good to excellent tires, dry surface.  A reasonably skilled driver can stop at 20 ft/sec² (0.62 g's).   Most production street vehicles have a maximum braking around 0.8 g's.

| | | | |
|---|---|---|---|
| BMW M3 | 1.0 g's | 32.3 ft/s² | 22.0 mph/sec |
| Toyota Celica GT | 0.94 g's | 30.2 ft/s² | 20.6 mph/sec |
| Lincoln Continental | 0.92 g's | 29.6 ft/s² | 20.2 mph/sec |
| Nissan Maxima | 0.85 g's | 27.3 ft/s² | 18.6 mph/sec |
| Chevy Blazer | 0.76 g's | 24.5 ft/s² | 16.7 mph/sec |
| Dodge Colt GL | 0.72 g's | 23.2 ft/s² | 15.8 mph/sec |

Table A-4 -- **Maximum Braking for select Vehicles**
Productions years 1991 - 1995.  Braking to a stop
measured on a dry flat surface starting at 60 mph.

## General Equations

$$v = v_0 + a\,t$$

$$d = v_0\,t + 0.5\,a\,t^2$$

t = time                                    a = acceleration (+a) or deceleration (-a)
v = speed at time t                    $v_0$ = speed at time t = 0.
d = distance traveled at time t

NOTES

# Appendix B -- Constants/ Conversions

## Multiplier Prefixes

| Multiplier | Power | Prefix | Symbol | |
|---|---|---|---|---|
| | $10^{100}$ | | | googol |
| 1,000,000,000,000 | $10^{12}$ | tera | T | trillion |
| 1,000,000,000 | $10^{9}$ | giga | G | billion |
| 1,000,000 | $10^{6}$ | mega | M | million |
| 1,000,000 | $10^{3}$ | kilo | k | thousand |
| 100 | $10^{2}$ | hecto | h | |
| 10 | 10 | deka | da | |
| 0.1 | $10^{-1}$ | deci | d | |
| 0.01 | $10^{-2}$ | centi | c | |
| 0.001 | $10^{-3}$ | milli | m | |
| 0.000 001 | $10^{-6}$ | micro | $\mu$ | |
| 0.000 000 001 | $10^{-9}$ | nano | n | |
| 0.000 000 000 001 | $10^{-12}$ | pico | p | |
| 0.000 000 000 000 001 | $10^{-15}$ | femto | f | |
| 0.000 000 000 000 000 001 | $10^{-18}$ | atto | a | |

In the U.S. sometimes g's or a grand refers to a thousand (1,000).
In some European countries $10^{18}$ is a trillion, and $10^{12}$ is a billion.

## Speed of Light (c) in a Vacuum

| | |
|---|---|
| 299,792,456.2 | meters / second |
| 983,571,050.5 | feet / second |
| 1,079,252,842 | kilometer / hour |
| 670,616,625.4 | miles / hour |
| 582,749,914.9 | nautical miles / hour |
| 299,792.4562 | kilometers / second |
| 186,282.3959 | miles / second |
| 161,874.9763 | nautical miles / second |
| 983.571051 | feet / microsecond |
| 0.983 571 051 | feet / nanosecond |

## Speed of Sound (Mach 1)

| Sea Level | | 36,000 - 82,000 feet | | 82k to 154k feet |
|---|---|---|---|---|
| 765 mph | Linear decrease to | 660 mph | Linear Increase to | 755 mph |

Speed of Sound at Sea Level: 1 mile / 4.7 seconds
Between 36,000 - 82,000 feet: 1 mile / 5.5 seconds

## Other Constants

| Permittivity of free space: | $e_0 = \mu_0^{-1} c^{-2}$<br>$e_0 = 8.854\ 185 \times 10^{-12}$ | Farads / meter |
|---|---|---|
| Permeability of free space: | $\mu_0 = 4\ \pi \times 10^{-12}$<br>$\mu_0 = 1.266\ 370\ 61 \times 10^{-6}$ | Henrys / meter |
| Intrinsic Impedance of free space: | $Z = 376.73$ | ohms |
| Acceleration due to Gravity: | $g = 9.80665$<br>$g = 32.1740$ | meters / second second<br>feet / second second |
| Boltzmann's Constant | $k = 1.380\ 622 \times 10^{-23}$ | watt seconds / °Kelvin |

## Transcendental Numbers

Base of Natural Logarithms
e = 2.71828 18284 59045 23536 02874 71352 66249 77572 47093 69995...

pi
π = 3.14159 26535 89793 23846 26433 83279 50288 41971 69399 37510...
π radians = 180°
1 radian = 180°/π = 57.3°

## Typical U.S. Interstate

| Left<br>Shoulder | Lane<br>Width | Right<br>Shoulder | Height Clearance | Maximum Grade |
|---|---|---|---|---|
| 3 feet | 12 feet | 10 feet | 14 feet | 5% City<br>6% Mountains |

Standard Railroad Gauge: 4 feet, 8.5 inches

## Speed Conversions

| | | | |
|---|---|---|---|
| 1 meter / second = | 1 / 0.3048 | feet / second | 3.28 ft/s |
| | 3.6 | kilometers / hour | |
| | 1 / 0.44704 | mph | 2.24 mph |
| | 900 / 463 | knots | 1.94 knots |
| 1 feet / second = | 0.3048 | meters / second | |
| | 1.09728 | kilometers / hour | |
| | 15 / 22 | mph | 0.68 mph |
| | 274.32 / 463 | knots | 0.59 knots |
| 1 kilometer / hour = | 1 / 3.6 | meters / second | 0.28 m/s |
| | 5 / 5.4864 | feet / second | 0.91 ft/s |
| | 1 / 1.609344 | mph | 0.62 mph |
| | 1 / 1.852 | knots | 0.54 knots |
| 1 mph = | 0.44704 | meters / second | |
| | 22 / 15 | feet / second | 1.47 ft/s |
| | 1.609344 | kilometers / hour | |
| | 402.336 / 463 | knots | 0.87 knots |
| 1 knot = | 463 / 900 | meters / second | 0.51 m/s |
| | 463 / 274.32 | feet / second | 1.69 ft/s |
| | 1.852 | kilometers / hour | |
| | 463 / 402.336 | mph | 1.15 mph |

knot = nautical miles / hour

## Distance Conversions

| | | |
|---|---|---|
| 1 inch = | 0.0254 | meters |

| | | |
|---|---|---|
| 1 feet = | 12 | inches |
| | 1 /3 | yards |
| | 0.3048 | meters |

| | | |
|---|---|---|
| 1 yard = | 36 | inches |
| | 3 | feet |
| | 0.9144 | meters |

| | | | |
|---|---|---|---|
| 1 meter = | 1 / 0.0254 | inches | 39.37 in |
| | 1 / 0.3048 | feet | 3.28 ft |
| | 1 / 0.9144 | yards | 1.09 yds |

| | | | |
|---|---|---|---|
| 1 kilometer = | 1 / 1.852 | nautical miles | 0.54 NM |
| | 1 / 1.609344 | miles | 0.62 mi |
| | 1 / 0.0009144 | yards | 1094 yds |
| | 1 / 0.0003048 | feet | 3281 ft |
| | 1000 | meters | |

| | | | |
|---|---|---|---|
| 1 mile (statute) = | 1609.344 / 1852 | nautical miles | 0.87 NM |
| | 1.609344 | kilometers | 1.61 km |
| | 1760 | yards | |
| | 5280 | feet | |
| | 1609.344 | meters | |

| | | | |
|---|---|---|---|
| 1 nautical mile = | 1852 / 1609.334 | miles | 1.15 mi |
| | 1.852 | kilometers | |
| | 1852 / 0.9144 | yards | 2025 yds |
| | 1852 / 0.3048 | feet | 6076 ft |
| | 1852 | meters | |

| | | |
|---|---|---|
| 1 nautical mile (UK) = | 1853.184 | meters |

# Decibel (dB)

$$\text{ratio} = 10^{dB/10} \qquad\qquad dB = 10\log(\text{ratio})$$

$$mW = 10^{dBm/10} \qquad\qquad dbm = 10\log(mW)$$

$$W = 10^{dBW/10} \qquad\qquad dBW = 10\log(W)$$

W = Watts
mW = milliwatts

+3 dB = 2.0          dB + dB = dB
-3 dB = 0.5          dBm + dB = dBm
                     dBm + dBm = nonsense

| RANGE | RADIO | RADAR |
|-------|-------|-------|
| double | +6 dB | +12 dB |
| half | -6 dB | -12 dB |

# Receiver Noise Power In
## $N_i = kTB$

$N_i$ = Noise Power In (watts)
k = Boltzmann's Constant = 1.380 622 x 10$^{-23}$ watt sec / °K
T = Temperature in Degrees Kelvin (K°)
B = Receiver Noise Bandwidth in Hertz (Hz)

### $kTB = -174$ dBm/Hz

T = 300°K (80°F), B = 1 Hz

*Isotropic Antenna Gain* (dBi) compared to
*Standard Dipole Antenna Gain* (dBD)

$$dBi = dBD + 2.15$$
$$dBD = dBi - 2.15$$

# Additional Reading / References

## DATA REFERENCES

1. APCO Code of Practice for Operational Use of Road Policing Enforcement Technology,ACPO 2004 version 2.3, Association of Chief Police Officers (England, Whales & N Ireland), 25 Nov 2004.

2. American National Standard Safety Levels with Respect to Human Exposure to Radio Frequency Electromagnetic Fields (300 kHz - 100 GHz), ANSI Standard C95.1- 1982.

3. Annual Reference Guides, 1999 and 1997, Compliance Engineering

4. CRC Standard Mathematical Tables, 20th Edition, Samuel M. Selby, CRC Press, 1972.

5. Driver Reaction Time in Crash Avoidance Research: Validation of a Driving Simulator Study on a Test Track IEA2000 ABS IDS VRTC, Daniel V. McGehee, 2000.

6. International System of Units (SI), Physical Constants and Conversion factors, 2nd revision, E. A. Mechtly - University of Illinois, NASA SP-7012, 1973/

## TEXT BOOKS

1. Antenna Engineering Handbook, 1st Edition, Henry Jasik, McGraw-Hill, 1961

2. Digital Signal Processing, Alan V. Oppenheim ,Ronald W. Schafer, Prentice-Hall, 1975.

3. Fundamentals of Waves, Optics and Modern Physics, 2nd Edition, Hugh D. Young, McGraw-Hill, 1976.

4. Introduction to AIRBORNE RADAR, George W. Stimson, Hughes Aircraft Co. - Radar Systems Group, CA, 1983.

5. Introduction to Radar Systems, 2nd Edition, Merrill I. Skolnik, McGraw-Hill, 1980.

6. Radar Handbook, Merrill I. Skolnik, McGraw-Hill, 1970.

## POLICE RADAR BOOKS

1. Beating The Radar Rap, 2nd Edition, Dale T. Smith & John Tomerlin, Bonus Books Inc., Chicago, 1990.

2. Case Dismissed II, Radio Association Defending Airwave Rights, Inc. (R.A.D.A.R.), 1997.

3. Speed Monitoring Technology Handbook, Radio Association Defending Airwave Rights, Inc. (R.A.D.A.R.), 1996.

## U.S. GOVERNMENT DOCUMENTS

1. Calibration Techniques for Across-the-Road Traffic Radars, National Institute of Standards and Technology, Radio-Frequency Technology Division, Claude M. Weil, NIST Technical Note 1398, May 1998.

2. FCC Release: First Report and Order (FCC 02-211). Commission Requires Radar Detectors to Comply with Emission Limits to Prevent interference to Satellite Services, Action by the Commission July 12, 2002.

3. FCC Public Notice -- FCC Regulates Radar Transmitters, but not Radar Detectors, Delegated Authority DA 96-2040, 1996 DEC 09.

4. FCC Rules and Regulations, Parts 15, 20, and 90.

5. Occupational Exposure of Police Officers to Microwave Radiation from Traffic Radar Devices, U.S. Department of Labor (osha), W. Gregory Lotz, Robert A. Rinsky, Richard D. Edwards, National Technical Information Service (NTIS) publication number PB95-261350, June, 1995.

6. Performance Standards for Speed Measuring Devices, United States Department of Transportation and National Highway Traffic Safety Administration, Federal Register Volume 46, Number 5, January 8, 1981.

7. Police Traffic Radar Issue Paper, United States Department of Transportation and National Highway Traffic Safety Administration, DOT HS-805 254, February 1980.

# Index

www.ingramcontent.com/pod-product-compliance
Lightning Source LLC
Chambersburg PA
CBHW081119170526
45165CB00008B/2493